I0222846

REFRACTED ECONOMIES

Diamond Mining and Social Reproduction in the North

Rebecca Jane Hall

Since the beginning of the twenty-first century, diamonds have been lauded as a "glistening" driver of the northern Canadian economy. Canadian diamonds are cast with an imagined purity as though they had emerged by magic. However, these diamonds are mined on Dene land and extracted by people who fly in from afar, separated from their families for long periods of time.

Adopting a decolonizing and feminist approach to political economy, *Refracted Economies* analyses the impact of diamond mining in Yellowknife, Northwest Territories. The book centres on Indigenous women's social reproduction labour – both at the mine sites and at sites of community, home, and care – as a means of understanding the diffuse impacts of the diamond mines. Grounded in ethnographic work, the narratives of northern Indigenous women's multiple labours offer unique insight into the gendered ways northern land and livelihoods have been restructured by the diamond industry.

Rebecca Jane Hall draws on documentary analysis, interviews, and talking circles in order to understand and appreciate the – often unseen – labour performed by Indigenous women. Placing this day-to-day labour at the heart of her analysis, Hall shows that it both reproduces the mixed economy and resists the gendered violence of settler colonialism as exemplified by extractive capitalism.

REBECCA JANE HALL is an assistant professor in the Department of Global Development Studies at Queen's University.

Refracted Economies

Diamond Mining and Social Reproduction in the North

REBECCA JANE HALL

UNIVERSITY OF TORONTO PRESS
Toronto Buffalo London

© University of Toronto Press 2022
Toronto Buffalo London
utorontopress.com

ISBN 978-1-4875-4083-8 (cloth)   ISBN 978-1-4875-4086-9 (EPUB)
ISBN 978-1-4875-4084-5 (paper)   ISBN 978-1-4875-4085-2 (UPDF)

Library and Archives Canada Cataloguing in Publication

Title: Refracted economies : diamond mining and social reproduction in the
   North / Rebecca Jane Hall.
Names: Hall, Rebecca Jane, author.
Description: Includes bibliographical references and index.
Identifiers: Canadiana (print) 20210335386 | Canadiana (ebook) 20210335475 |
   ISBN 9781487540845 (paper) | ISBN 9781487540838 (cloth) | ISBN
   9781487540869 (EPUB) | ISBN 9781487540852 (PDF)
Subjects: LCSH: Diamond mines and mining – Social aspects – Northwest
   Territories – Yellowknife. | LCSH: Diamond mines and mining – Economic
   aspects – Northwest Territories – Yellowknife. | LCSH: Indigenous women –
   Northwest Territories – Yellowknife – Economic conditions. | LCSH:
   Indigenous women – Northwest Territories – Yellowknife – Social conditions.
   | LCSH: Yellowknife (N.W.T.) – Economic conditions. | LCSH: Yellowknife
   (N.W.T.) – Social conditions.
Classification: LCC TN994.C3 H35 2022 | DDC 338.2/782097193–dc23

We wish to acknowledge the land on which the University of Toronto Press
operates. This land is the traditional territory of the Wendat, the Anishnaabeg,
the Haudenosaunee, the Métis, and the Mississaugas of the Credit First
Nation.

University of Toronto Press acknowledges the financial support of the
Government of Canada, the Canada Council for the Arts, and the Ontario Arts
Council, an agency of the Government of Ontario, for its publishing activities.

Canada Council   Conseil des Arts
for the Arts   du Canada

ONTARIO ARTS COUNCIL
CONSEIL DES ARTS DE L'ONTARIO
an Ontario government agency
un organisme du gouvernement de l'Ontario

Funded by the   Financé par le
Government   gouvernement
of Canada   du Canada

Canada

For Nico.

Contents

List of Images and Figures   ix

Acknowledgments   xi

1 Introduction   3

Part One: Theorizing the Northern Mixed Economy

2 An Expanded Approach to Production   27
3 Wıìlıìdeh's Mixed Economy   48

Part Two: The Political Economy of Diamonds

4 The Global Political Economy of Canadian Diamonds   79
5 The NWT Diamond-Mining Regime   96

Part Three: Indigenous Women's Labour and the Diamond Mines

6 Time, Place, and the Diamond-Mining Regime   123
7 Social Reproduction and the Diamond-Mining Regime   151
8 Diamonds, Subsistence, and Resistance   182

9 Conclusion   215

Appendix   229

Notes   231

References   247

Index   263

Images and Figures

Images

Image 1   Photo taken by author beside Yellowknife City Hall.   6
Image 2   Giant Mine on the shores of Great Slave Lake.   61
Image 3   NWT Diamond Mines.   102
Image 4   Diavik Diamond Mine.   216

Figures

Figure 1   An expanded conception of production.   32
Figure 2   Percentage (%) of territorial GDP by industry (2019).   130
Figure 3   Employment, by Class of Worker. Northwest Territories, 2001–2019.   130
Figure 4   Rates of employment.   133
Figure 5   Percentage of persons 15 and over who trapped in a given year, by region, Northwest Territories, 1988–2018.   188
Figure 6   Percentage of persons 15 and over who hunted or fished in the year, by community, Northwest Territories, 1998–2018.   189
Figure 7   Percentage of households where 75% or more (most or all) of the meat or fish eaten in the household, by community, Northwest Territories, 1998–2018.   190

Acknowledgments

This book is the outgrowth of years of generous collaborations and conversations spanning kitchen tables, seminars, and community gatherings. I am so grateful to all those whose work and thinking has changed and enlivened mine.

Thank you to the Native Women's Association of the Northwest Territories staff and community for welcoming me as a staff member, and, later, for hosting and contributing so much to my research. Thanks to Della Green and Jennifer Hunt-Poitras for all of your intellectual guidance and logistical support. Special thanks to Marie Speakman, my mentor, friend, and colleague, who demonstrates daily the strength and wisdom of compassion. Thank you, also, to the many community activists and organizations who guided me in this research, offering insight, encouragement, and organizational support, including Arlene Hache, Lesley Johnson, Lois Little, Nancy Macneill, Lawrence Nahtene, Lorraine Phaneuf, Itoah Scott-Enns, Anneka Westergreen, Alternatives North, the Centre for Northern Families, Ecology North, the Status of Women Council of the NWT, the YWCA, the NWT and Nunavut Chamber of Mines, and especially all of the participants in interviews and talking circles. More recently, I've had the opportunity of collaborating with Hotıì ts'eeda, the Tłı̨chǫ Dene First Nation, and the Yellowknives Dene First Nation. These collaborations have done much to enrich my thinking as I have developed this manuscript, and I am grateful for the generosity of my Dene and northern hosts. Thank you to Lena Black, Tee Lim, Rachel Macneill, Jessica Simpson, Tyanna Steinwand, and John B. Zoe.

Throughout this work, I have been fortunate to receive guidance and friendship from a number of mentors. The brilliant and thoughtful Leah Vosko has taught me so much through her kind attention to my work and by example. Frances Abele, Kamala Kempadoo, Elaine

Coburn, Ethel Tungohan, and Abigail Bakan have all given wonderful feedback and support that have helped to shape the book as it now stands. Thank you, also, to Rawwida Baksh, Andrea Doucet, Susan Ferguson, Rianne Mahon, and David McNally for their intellectual and personal guidance over the years. The thinking in this book has been shaped by my participation in a number of conferences and journal publications. In particular, many thanks to the organizers of and participants in the 2014 Historical Materialism London panels on social reproduction and subsequent special issue; and to the organizers of and participants in the 2015 Consuming Intimacies symposium (Brock University) and subsequent special issue (Studies in Social Justice). Thank you to Ezgi Dogru, Jessica Evans, Simon Granovsky-Larsen, Tobin Leblanc Haley, Heather McLean, and Amanda Wilson for travelling this academic journey with me.

The Department of Global Development Studies at Queen's University has provided a stimulating and nurturing environment from which to prepare this manuscript. I am grateful to my colleagues and the administrative staff in the department for their comradely support and for their everyday examples of social justice scholarship. I have been fortunate to take up my position at Queen's alongside an inspiring cohort of new faculty and in the company of truly wonderful graduate and undergraduate students. I'm very grateful for the many conversations that have sharpened my thinking in this text, including those with Diana Córdoba, Alexandria Knipp, Reena Kukreja, Carolyn Prouse, Susanne Soederberg, Marcus Taylor, and Kyla Tienhaara.

Thank you to the folks at University of Toronto Press for their work in preparing this manuscript. Daniel Quinlan, my editor at University of Toronto Press, has brought his generosity and keen eye to shepherding this project. This book was tremendously improved through the suggestions of the anonymous reviewers and the University of Toronto Press board. Warm thanks to Pat Kane for the donation of his beautiful photograph, seen on page 216. Martha Hall, Don Hall, Bridget Hall, Amanda Wilson, Tobin LeBlanc Haley, and Nico Koenig all gave feedback on various iterations of this manuscript and made it better. Special thanks to Madeline Hall for taking on the whole book. Hannah Ascough and Brandon Pryce offered great help in the final preparations of the manuscript. Nearing the end of this process, I got stuck on a title and turned to friends and family members for help. I need to write ten more books on diamond mines to do justice to your creative titles. Warm thanks to Brendon Goodmurphy and Jill Dalton for the ideas that made the cut.

Finally, and especially, thank you to my warm and wonderful circles of friends and extended family. I could not have done this without you and I wouldn't have wanted to. My parents and grandparents, Martha Hall, Don Hall, Jane McKeague, Alex McKeague, Rachel Hall, Ross Hall, and Anne Hall, are examples of lives lived deliberately, with intelligence and with love. Mom, Dad, Madeline, and Bridget, you are radically supportive and have been a tremendous source of love and encouragement during this research and writing, and always. Dad, though I tease about the terror provoked by your editing pen, I am so grateful for the time you've taken with my writing over the years. To my partner, Nico Koenig, thank you for everything. Our daily conversations about social reproduction have helped to shape this book, and your extra household labours have made its writing possible. I gave birth to both my children while working on this book. Hattie and Fern, thank you for pushing me to write quickly because I'd always rather be hanging out with you, and to write well because, if I'm not hanging out with you, I'd better be doing something halfway decent.

This research was funded by the Social Sciences and Humanities Research Council, the Ontario Graduate Scholarship, York University, and Queen's University.

REFRACTED ECONOMIES

1 Introduction

For many settlers, the land north of the 60th parallel has long been a repository of resources and dreams. The "strange things done in the midnight sun" captured European imaginations and came to represent an intrepid "Canadian-ness," while the "men who moil for gold" shaped a national political economy built upon resources extracted from land imagined as empty. In emptying the "Arctic trails"[1] of their intricate Indigenous socio-economies with histories stretching back to time immemorial, settlers were attempting to make of the North what they needed. The rhetorical and material emptying of land is a settler colonial strategy linking disparate times and places around the globe, shifting in its specificities and its specific violences. In northern Canada, the frontier narrative of empty land has evolved from the explicit to the implicit but perseveres nonetheless.[2] The Northwest Territories (NWT) capital, Yellowknife, has grown from a gold-mining town to the territory's administrative centre, with tall office towers housing federal, territorial, and municipal government bureaucracies as well as the corporate interests that manage the diamond mines operating a few hundred kilometres northwest of the city. Many of the shacks first erected by gold miners in the 1930s remain, but these sit alongside fresh condominiums and suburban developments that might have been air-dropped from any number of small southern Canadian towns. Notwithstanding the myriad restaurants flooded daily at 11:55 a.m. by government workers on their lunch break and the fitness centres filled with residents escaping the winter darkness, the frontier rhetoric persists, and it shapes the narrative of Canadian diamonds, the subject of this book.[3]

While the frontier imaginary offers evocative fodder for engagement and critique (Sabin 2014), it belies the Dene, Inuit, and Métis history of the land, a history that stretches so far back that settlers' brief engagement in the North seems almost inconsequential. Almost. The frontier imaginary denies the past and present processes of settler colonialism that have established extractive industries in the North, the reliance on Indigenous expertise in their establishment, and the ongoing Indigenous resistance to the encroachment of extractive political economies. In the frontier account, white settler labour carved a resource economy out of wild and rough terrain. The colonial violence that exploited peoples and resources has been reframed as perseverance in a hostile environment, while subsistence socio-economies are trapped outside of history, in anachronistic space.[4]

In opposition to the frontier imaginary, a very real social alternative to Western capitalism plays out in the day-to-day labours of people in the NWT (and throughout the North): the mixed economy. By "mixed economy" I mean a regional political economy that combines Indigenous and settler modes of life. Here I am drawing upon Glen Coulthard's expanded approach to social relations (Coulthard 2014); by modes of life, I mean ways of producing and reproducing, ways of organizing, and ways of knowing, interpreting, and communicating the world. In the North's mixed economy, capitalist production has developed alongside and in tension with land-based subsistence practices and relations oriented toward communal needs. Subsistence in this region is practised through what Coulthard calls grounded normativity: "the modalities of Indigenous land-connected practices and longstanding experiential knowledge that inform and structure our ethical engagements with the world and our relationships with human and nonhuman others over time" (2014: 13). Thus, over time, the ever-shifting social relations of this place have borne the tensions – gendered, racialized, and, arguably, violent – between the capitalist temporal imperative to extract surplus value and subsistence-oriented relations revolving around peoples, land, and animals.[5] This mixed economy is reproduced through long-standing and resilient Indigenous modes of life, but one cannot assume this will always be so. The mixed economy and its constitutive parts amount to a shifting social formation – and, given capital's attributes of accumulation and annexation, a possibly temporary one. Far from being a fixed socio-economic structure, the mixed economy is the dynamic result of ongoing Indigenous resistance to colonization. The sustained and transformative labours of the Inuit, Dene, and Métis socially reproduce Indigenous modes of life in the face of political and economic impositions that threaten the North's distinctive land-based relations.

In the NWT, the most recent imposition on the mixed economy has been diamond mining. Within a strikingly few short years, diamond mining – a new commodity for Canada – has become the NWT's pre-dominant industry. The first diamond mine opened in the NWT in 1998. Since then, Canada has become the third-largest diamond pro-ducer in the world. The industry now accounts for as much as half the NWT gross domestic product (GDP) (GNWT 2018a). Glossy magazine ads celebrate the "purity" of Canada's northern diamonds by depict-ing vast, beautiful, and empty landscapes, thus obscuring the complex social relations and associated social problems the diamond-mining re-gime has generated. These social problems include some of Canada's highest rates of violence against women. It was, in fact, the anecdotal and observed relationship between the diamond mines and violence against northern women, especially Indigenous women, that led me to this research. I am a white settler, raised in Ottawa, Ontario, with Irish, English, and Scottish ancestry. I moved to Yellowknife in 2008 and took up a position doing front-line and advocacy anti-violence work at the Native Women's Association of the NWT. I spent my days working for their Victim Services Program with women who have experienced violence. The pervasiveness of this violence demanded and continues to demand an informed response. Too often, gender-based violence, sexual violence, and intimate partner violence in marginalized com-munities is naturalized, approached as a tragic, but inevitable and somewhat inexplicable, aberration from the safe, non-violent norm. But violence is *not* normal, and I wondered what this violence might be saying – screaming – about the changing political economy of the region and its gendered and colonial contours.

Anyone who starts looking into the NWT's contemporary social, po-litical, and economic landscape will immediately bump into the dia-mond industry, and the mines quickly became my subject of inquiry. The labour force in Yellowknife is remarkably diverse for a town of 20,000, as are the ways of life carved out by its residents. Most sala-ried jobs are in public administration (First Nation, municipal, territo-rial, and federal), but the city is also characterized by a vibrant social economy (Southcott 2015), as well as an arts industry and a growing service industry. All of this employment operates alongside ongo-ing land-based, or subsistence, production and distribution, which includes gathering, fishing, hunting, trapping, and the preparation of plant and animal materials (for food, medicine, clothing, art, and household materials), for consumption, trade, or sale. Yet for all this diversity, there is a ubiquity to the presence of the diamond-mining regime. "Diamond-mining regime" here refers to the policies, laws, and norms that have been developed by the Canadian state, private capital, and Indigenous communities. In the town centre, flags fly over Sombe

Image 1. Photo taken by author beside Yellowknife City Hall.

K'e (a park and meeting space adjacent to City Hall) and Franklin Street (the main drag), declaring Yellowknife the "Diamond Capital of North America" (see Image 1), while diamond company logos adorn office buildings and banners around the town. Kin, communities, and the service industry move to the tempo of two-weeks-on/two-weeks-off, the most common schedule for the mine workers.

The diamond mines have been carved into the boundary between the accumulation of capital (wherein land is understood as extractable resources) and the place-based relations of Indigenous communities (wherein land is understood through its reciprocal relations with people and animals (Blondin 1997; Legat 2012) that characterizes the northern mixed economy. Thus, the main question this book asks is how the diamond-mining regime has impacted what Coulthard (2014) refers to as the "delicate balance" of the mixed economy. Drawing on insights from Indigenous and anti-racist feminists, and rooted in a feminist political economy (FPE) approach that makes visible the inextricable relation between social reproduction and production-for-surplus (or profit), I focus on the gendered aspect of this impact, orienting my analysis around Indigenous women's labours and experiences. I suggest that Indigenous women's central role in protecting and reproducing the mixed economy has shaped their engagement with the diamond mines. Thus, the stories of their multiple labours – productive and reproductive, land-based and waged – offer unique insight into the racialized and gendered restructuring of NWT social relations imposed by the diamond-mining regime and the structural violence it has wrought. The specific narratives discussed in this book, in turn, are illustrative of the general violence of ongoing processes of settler capitalist dispossession, which have become more insidious but no less targeted over time. This study reveals the continuity across time and place in settler colonial assaults targeting Indigenous women and their reproductive capacities (social and biological), for what they reproduce is transformative and does not yield to the totalizing forces of state and capital.

I. Grounding the Mixed Economy

The region of study is Dene territory. It includes Yellowknife and the adjacent Wıìlıìdeh (Yellowknives Dene) towns of Dettah and N'dilo, as well as the Tłı̨chǫ Dene capital, Behchokǫ̀ (100 kilometres up the river from Yellowknife). The NWT, including the region of study, has also long been home to Inuit and Métis peoples.[6] While the Dene of the region have shared and intertwining histories that span the land that is now the NWT, communities operate through distinct political formations and identities.[7] The traditional land of the Yellowknives Dene includes the city of Yellowknife itself; however, as a result of European settlement in what is now Yellowknife and government efforts to establish permanent Dene settlements, the Wıìlıìdeh Yellowknives Dene now live primarily in N'dilo and Dettah. N'dilo sits on the tip of Latham Island, neighbour to Yellowknife's Old Town, looking out on Great Slave Lake – what Peter Kulchyski describes as "prime real estate ... poor people on expensive turf" (2005)[8] – while Dettah is around the bay, a 40-minute drive from Yellowknife in the summer or a ten-minute drive along the frozen Great Slave Lake during the winter. Having settled the Tłı̨chǫ Land Claims and Self-Government Agreement in 2003 (discussed below), the Tłı̨chǫ Dene First Nation now governs 39,000 square kilometres of land northwest of Yellowknife, surrounding the four Tłı̨chǫ communities of Whatì, Gamètì, Wekweètì, and Behchokǫ̀. Behchokǫ̀, 100 kilometres north of Yellowknife and the largest community of the four, is the governing centre of the Tłı̨chǫ First Nation. John B. Zoe describes the contemporary Tłı̨chǫ landscape, explaining how the land, and knowledge and ethics of the land, are woven into and protected by Tł chǫ language and culture: "Known intimately to Tł chǫ Elders, trails, which are used year-round, provide access to a vast harvesting region, and link thousands of place names, each with a narrative of some form, sometimes many, inextricably bound to the place. Names and narratives convey knowledge, and in this way Tł chǫ culture is tied directly to the landscape" (Zoe 2014).

Like Indigenous peoples from other territories and communities, some Yellowknives Dene and Tł chǫ Dene live in the city of Yellowknife itself. Yellowknife's name comes from the Wıìlıìdeh Yellowknives Dene, named for their tools and cooking instruments, which they made from copper they collected from their traditional territory (YK Dene First Nation Advisory Council 1997). Today, Yellowknife is NWT's capital and its centre of private capital and government activities. People from southern Canada – indeed, from around the globe – have come to Yellowknife, having been recruited into the extractive and auxiliary industries, including through state initiatives such as the temporary foreign worker program.[9] Yellowknife has also become home (in varying degrees of "permanence") to Indigenous people from across the NWT and the North. Indeed, during a quick stroll along Franklin Avenue – the town's main road – one might encounter a conversation in Amharic or, alternatively, Inuktitut, followed by a group of Tł chǫ Dene youth injecting Dogrib into their English banter.[10]

Yellowknife, N'dilo, Dettah, and Behchokǫ̀ are discrete sites, each with its own distinct dynamics of community-level social reproduction; but they are also tightly linked through border-crossing social relations. For example, some residents live in Behchokǫ̀ and work in Yellowknife, or work in Yellowknife and live in Behchokǫ̀, driving one hour each day along the road that links the two towns. Family relations tightly link the communities to Yellowknife residents. Rose, a young Tł chǫ Dene woman who lives in Yellowknife, said: "Yeah, we go back every two months or so. Because I still have cousins out there. And even my immediate family, like it's my mom and my dad and my sister, and me growing up with millions of cousins running around. I want my son to have that ... And because the world's so different now, we go out there. And he has a sense of freedom there" (Interview 102). Many Tł chǫ conceptualize their lives, families, and communities as straddling these two spaces.

Thus, the mixed economies of northern cities like Yellowknife should be understood in terms of their relationship to a rural base (Abele 2009a) – that is, to the smaller communities more deeply and consistently engaged in subsistence, as well as to the expanses of land and water where the caribou live and where the trap lines and fishnets are set. This is not to say that people do not engage in subsistence production in Yellowknife and the other urban centres of the NWT, or that capitalist production is not important to the economy in smaller communities; rather, the balance between capitalist and subsistence production is significantly different in urban and rural communities, and the urban and the rural in the NWT are linked through a social and economic interdependence. Taken together, the cities and communities represent a mixed economy, one in which livelihoods are built through a combination of subsistence, social reproduction, and wage labour in government (municipal, territorial, federal), non-governmental organizations, and private business (mining and industries auxiliary to both mining and government, and tourism, as well as businesses serving the local populations).

Traditionally, however, the land on which Yellowknife is situated was inhabited primarily by the Wıı̀lıı̀deh Yellowknives Dene, whose summer camps lined the Yellowknife River, and the Tłı̨chǫ, whose summer camps were just to the north, up the northern arm of Great Slave Lake. While it is accurate – and indeed important – to name these spaces as the traditional territory of the Dene, the concept "traditional" must not be conflated with a static account of history. As with any group or community, land use has changed over time. Dene history chronicles these changes through relationships: relationships to people (outside and inside Dene communities), to animals, and to land. Gibson MacDonald and colleagues (2014) write that "Tłı̨chǫ history, when orated by elders and leaders, is noticeably oriented around five central agreements and exchange relationships (that are roughly cast into five eras of history starting with prehistory or the formation of the Tłı̨chǫ cosmology) with animals, other aboriginal groups, the federal government, and mining companies" (65). John B. Zoe (in Gibson 2008) explains that eras are marked "in the initiation of relationships to newcomers, the negotiation of difference and the resolution of difference through agreement-making" (53). For example, the second historical era for the Tłı̨chǫ tells the story of the 1823 peace treaty between the Tłı̨chǫ and the Wıı̀lıı̀deh Yellowknives Dene. After years of hostility generated by competition for European goods, the two groups struck a peace treaty that demarcated their respective lands. That treaty made it possible for the Tłı̨chǫ to camp closer to the Yellowknife River and facilitated the sharing of land, and it continues to shape competing interpretations of, and struggles for, the land (YK First Nation Advisory Council 1997; Gibson 2008: 67).

Thus, Tłı̨chǫ cosmology is oriented around ruptures in social relations that are then negotiated and resolved. The concept of "rupture" acknowledges the impact of new relations and the labours necessary to manage that impact, but it does not approach the new relations as totalizing. Instead, eras marked by the rupture of new relations make space for a new set of dialectically emergent relations. What is distinctive about the Tłı̨chǫ approach to new eras/relationships is the importance it places on the past and on the values that were established in previous eras when new agreements and practices and the new relationships that came with them were being forged. This way of approaching history is a reminder of the agency of northern Indigenous peoples in shaping the social relations of the mixed economy; it also reminds us that, while the mixed economy is evidence of Indigenous resistance and regional innovations, it is neither fixed nor a given. Rather, the mixed economy will continue to be shaped by new ruptures to come. It is not exempt from the threats of capital's *ongoing* tendencies toward new accumulations or from the structural violence of white supremacy. However, the concept of rupture – and its uneven spatial enactment across the urban and rural North – helps move analysis beyond the blunt instrument of dichotomous thinking (for example, traditional *or* capitalist) toward a grounded qualitative approach to the region's social relations and the implications of conceptualizing the mixed economy as a shifting and potentially temporary site of de/colonizing struggle. Inherent to "rupture" as a way of understanding and building upon history is an emphasis on people's relationships to one another and to the land, exemplifying the ethics of "grounded normativity." Glen Coulthard and Leanne Betasamoksake Simpson write that

> grounded normativity teaches us how to live our lives in relation to other people and nonhuman life forms in a profoundly nonauthoritarian, nondominating, nonexplotative manner. Grounded normativity teaches us how to be in respectful diplomatic relationships with other Indigenous and non-Indigenous nations with whom we might share territorial responsibilities or common political or economic interests. Our relationship to the land itself generates the processes, practices, and knowledges that inform our political systems, and through which we practice solidarity. (2016: 244)

I draw upon the concept of rupture throughout this book to link past processes of colonial restructuring with the impact, or rupture, of the diamond mines, and to contextualize processes of colonization within the much longer-standing Dene relations of human and non-human care and respect.

II. Gendering Diamond Mining in the Mixed Economy

In its relationship to northern Indigenous peoples and their labour, the diamond-mining regime represents both a continuity and a discontinuity with past forms of extraction. Gold mining came to the Yellowknife region in the 1930s, and the gold mines operated until the early 1990s. As another iteration of resource extraction driven by the imperatives of private capital and the Canadian state, the diamond-mining regime followed on the heels of the gold mines, sustaining the tension between subsistence – wherein land-based activities were oriented toward the daily and intergenerational well-being of the collective rather than the profit of the individual – and the extraction of surplus value. However, the diamond mines also represent something novel in terms of the mining regime's approach to relations with the surrounding Indigenous communities. The gold-mining industry was managed largely by settlers from southern Canada, but it relied on Indigenous expertise in establishing the mines, and it hired northern Indigenous people for specific tasks – mostly "bushwork" for men and cooking for women. At the time, however, the recruitment of Indigenous people for long-term mining employment was not a priority for either the gold companies or the Canadian state (Collymore 1980). By contrast, after decades of Indigenous organizing at the local, regional, and national levels, the diamond mines have taken a new approach to extraction, one that recognizes Indigenous relationships to the land (at least on paper) and that solicits Indigenous participation in the form of employment opportunities and consultations. Indigenous people in the NWT are now seen as potential mine workers, both as individual workers and through the contracting of Indigenous companies for mine operations. While the numbers vary from year to year, northern Indigenous hires today make up around 25 per cent of the mining workforce (GNWT 2018a); another 25 per cent come from the North, and the remaining 50 per cent are brought in from southern Canada or abroad.

The diamond mines also distinguish themselves from the regional mines of yesteryear in terms of how they organize production: namely, the fly-in/fly-out (FIFO) extractive model. FIFO is an increasingly popular approach to extraction that entails flying workers in for prolonged shifts (in contrast to traditional mining practice, wherein a mining town was built with all the necessary infrastructure to support its workers). FIFO enables mining companies to draw on workers who live at a distance from the mine, be they from southern Canada or one of the many communities of the NWT. A FIFO mining schedule requires workers to live at the mine site for their shifts, which are most often fourteen days, followed by fourteen days for travel home and – ostensibly – rest. Most daily shifts are twelve hours. All of this means that diamond mineworkers need not relocate (semi)permanently to the mine site and thus can maintain their homes outside the area of the mine. The ability to maintain one's home benefits both southern workers who do not want to live in the NWT and northern Indigenous workers who wish to remain in their home communities.

The diamond-mining regime's approach is part of a broader trend toward FIFO in extraction industries, both in Canada and internationally (Hann et al. 2014). This has contributed to the normalization of mobility as a quality of employment (Walsh 2012). The normalization of labour mobility is one expression of the sustained transience that has come to be expected in neoliberal industrial regimes around the globe (Vosko et al., eds, 2014: 4), a transience that is, in the Canadian context, particularly attached to the labour of resource extraction. More broadly, FIFO as it applies to diamond mining exemplifies neoliberal capitalism's emphasis on flexible and precarious labour, in that "the movement of transnational capital has also contributed to the institutionalization of temporariness as a permanently vulnerable condition for many people" (6). In commenting on the growing shift toward FIFO labour by extractive capital, Peck (2013) argues that this restructuring should be understood as more than a simple shift in the workplace balance of power, for it reflects "a much deeper social transformation. FIFO practices outsource the social-reproduction costs that the mining companies will no longer carry, and at the same time subvert the threat of workplace militancy, indeed of labour organization itself. They have individualized not only work but living conditions for the mining workforce" (237).

The individualization of labourers through the externalization of social reproduction from the site of production is key to the reorganization of social relations this research has identified. There is a fundamental tension – a fundamentally *gendered* tension – between the FIFO diamond-mining regime's intensification of individuation and the time-based drive for creating surplus value through the exploitation of labour, on the one hand, and the place-based, relational orientation that characterizes the labour that reproduces the NWT mixed economy, on the other.

When Indigenous people are approached as individual labourers to be recruited and retained, rather than as peoples enacting and grounded in distinct modes of life, Indigenous/settler relations come to be divorced from the very real tension between accumulation for surplus value and land-based subsistence relations. Having been externalized from the site of capitalist production, social reproduction bears much of the burden of this tension, for it is *this* labour that must manage the demands of the new extractive regime while also attending to gaps, or fissures, in subsistence relations resulting from that regime's impositions. Given Indigenous women's central role in daily and intergenerational social reproduction, the tension between the FIFO regime and subsistence falls with a particular intensity on their bodies and their labours. Racist ideologies that reframe subsistence as dematerialized culture or recreation and gendered ideologies that feminize and naturalize social reproduction deny this intensity and the violence associated with it. In this way, the FIFO diamond mining exemplifies the structural violence of reorganizing labour for the pursuit of capital accumulation, as well as the focus of this violence – racialized and gendered – upon the bodies of Indigenous women. This violence is neither natural nor inevitable but rather is rooted in the historical and contemporary materiality and ideology of capitalism and settler colonialism.

The observation that settler extractive projects import hierarchical Western gendered divisions of labour to the North is far from new; the fur trade, for example, imposed newly individualized and masculinized approaches to hunting and trapping (Bourgeault 1983) that, in some cases, impeded and devalued women's subsistence activities (Kuokkanen 2011). However, the spatial quality of FIFO extractive regimes – in that the site of (capitalist) production is physically divorced from the social reproduction of the community – has both obscured and facilitated the violent restructuring of social reproduction. FIFO demands workers who are unencumbered by care responsibilities, and this is articulated through patriarchal gender relations pervasive in Western capitalist societies, relations that are often magnified at sites of extraction (Scott 2007) and that construct capitalist production as masculinized. Conversely, social reproduction as the mirror image of this construction is both feminized and naturalized. Thus, the FIFO diamond-mining regime is a spatial articulation of the capitalist separation of production-for-surplus from social reproduction. In contrast to the interdependent modes of social reproduction performed across kin and community networks common to the mixed economy, the feminized responsibilities for social reproduction that result from the FIFO diamond-mining regime are individualized and localized at the household level, orienting resources away from responsibilities at the community and inter-household level. As a result of increased levels of Indigenous participation and the FIFO structure, the diamond-mining regime has exerted an intensified gendered pressure on place-based social relations in the mixed economy.

Social reproduction is a de/colonizing site of tension between capitalist accumulation and place-based relations. This brings a material imperative to violence against Indigenous women. However, activities linked to social reproduction can also be sites of resistance to the totalizing impulses of capitalism and the settler state. Throughout the book, I draw upon numerous examples offered by research participants of the day-to-day ways in which they seek to reproduce the unique place-based relations of their community: from prioritizing the needs of kin and community over the demands of the extractive regime, to intergenerational subsistence education. Indeed, the day-to-day labours that socially reproduce and strengthen a subsistence orientation in the face of the encroachment of capital and the Canadian state exemplify the ways in which the constrained choices made at the site of social reproduction shape the social relations that make up a given regional political economy and, in this context, reproduce the regional mixed economy. These are powerful acts of everyday decolonizing resistance. The processes of socially reproducing the mixed economy are difficult and sometimes messy and are, *contra* frontier narratives, the truest expression of northern "pluck," "bravery," and "imagination": the bravery, imagination, and ongoing creation of regional social relations that deny the persistent and crushing pressures of colonial capitalism. It is the tensions at the site of social reproduction performed by Indigenous women, and the decolonizing resistance and colonial violence operating at this site, that are at the heart of this book's inquiry.

IV. Social Location and Method

I engage in this research as a white settler who lives through the racialized privileges afforded to me by the white settler state and global racial hierarchies, and who aims to work in solidarity with and learn from the many sites of decolonizing struggle on Turtle Island and around the globe. I have lived my life thus far as an uninvited guest on Turtle Island, travelling across the Indigenous histories mapped onto this land. I grew up in Ottawa, the traditional and unceded territory of the Algonquin, and currently live in Kingston, the traditional and unceded territory of the Anishinaabe and the Haudenosaunee. I conducted this research with Dene, Métis, and Inuit communities on traditional and unceded Dene territory. For settlers working with Indigenous communities, locating oneself has become standard practice, but in becoming conventional, this practice must not lose its meaning. In locating myself, I ground and situate the analysis of this book in my history on this land, in the relationships that enrich and enliven the words herein, and in the blind spots that come with walking this world cloaked in white privilege. My writing is thus contained and explained by what I can and cannot know. But it is not excused, and neither am I: while the insights of this book emerged from the generous knowledge contributions of research partners, all errors and omissions are my own.

What, then, are the implications of my location for what this book is and is not? First and foremost, the knowledge contained on these pages is not mine. Research is never a possession, but it is worth being explicit about how much this is so. Formulating and posing the questions that guided this book was a collaborative and educative experience, and the analysis herein has been generated across conversations, across years. My experiences working in Yellowknife, and learning from friends, co-workers, and the community organizations with which I engaged between 2008 and 2010, led me to this research project. I developed research questions in conversation with the Native Women's Association of the NWT (NWA/NWT), community activists, and interviewees, asking what would be deemed useful and generative by the community. The NWA/NWT housed the project and provided invaluable support and expertise. They offered ongoing advice and insight as the fieldwork proceeded; they contacted their community networks to help recruit research participants; and they helped organize, design, and facilitate talking circles. The ongoing informal exchanges in the office as we discussed the research brought a rich reflexivity to the project that informed the evolving research design and analysis of the data.

For example, I had planned to conduct interviews with women who had been impacted by the diamond mines as well as focus groups with community workers. I thought that one-on-one interviews would be most appropriate for people discussing personal experiences, for I was concerned that discussions might involve sensitive matters that participants would want to keep confidential. However, staff at the NWA and other participating community groups highlighted the potential for community learning that could flow from hosting a talking circle. The talking circle is an Indigenous tool for bringing community members of all ages together for shared learning and listening (Wolf and Rickard 2003). The NWA has a history of using talking circles to discuss important and difficult community issues, most commonly issues related to residential schools. It was suggested that community members would feel more comfortable, and that discussion might be richer, if research was conducted through this method. So I coordinated two community talking circles; more than focus groups, these events were spaces where community members could come together to share their experiences of the diamond mines. The NWA staff and I cooked lunch for talking circle participants with the aim of creating a warm and inclusive atmosphere. Because talking circles are led according to Indigenous tradition and by a community leader, it was agreed that it would be appropriate for the talking circles to be led by Della Green, according to her traditions. Della is from Namgis First Nation, Alert Bay, and was the victim services coordinator at the NWA/NWT at the time. We developed questions to be displayed during the talking circle.[11] In talking circle tradition, participants are not called upon, but rather are given time to reflect and contribute when and if they desire. Displaying the research questions, rather than orally posing the questions consecutively, allowed the participants to respond to one another freely while personally reflecting on the questions and voicing their reflections at the time of their choosing.

The dialogue between participants led to rich insights and learning, as people were able to build on one another's reflections. Research participants shared tactics for managing hardships they had experienced and discussed community-building strategies to move beyond a socio-economy focused on extraction. Many participants expressed satisfaction and feelings of solidarity as they learned one another's stories; thus, the talking circles served the dual purpose of research and knowledge sharing. In this way, they helped orient this project away from knowledge "extraction" and toward dialogue. This is not a book *on* Indigenous peoples: this is a book written through settler/Indigenous relations about processes of settler colonialism. While I focus on Indigenous women's labour, I am not interested in delineating what it means to be a northern Indigenous woman.

What this book, does, instead, is develop one avenue toward understanding the ongoing ruptures in social relations in the mixed economy. In this way, through a feminist political economy lens, I endeavour to help reveal and challenge the power relations of which we are all a part. Informed by both the theoretical imperative to analyse the mixed economy by approaching subsistence, social reproduction, and capitalist production at the same level of analysis, and the methodological imperative to elevate the ethnographic authority of the research participants, I undertook a mixed-methods approach to data collection, combining historical analysis, documentary analysis, interviews, talking circles, and a focus group with community workers. While historical and contemporary documents provided data on the gold- and diamond-mining regimes, the policy apparatuses that surround them, and general socio-economic trends in the NWT, interviews, talking circles, and the focus group brought to the surface information about – often invisibilized – activities oriented toward subsistence and social reproduction performed by Indigenous women in the Yellowknife area.

I conducted fieldwork in the Yellowknife region (interviews, talking circles, the focus group, and document collection) from 1 June to 30 August 2014. In planning the fieldwork, I reached out to the NWA/NWT, my former employer, asking whether they wanted to be involved in the research, and if so, how. As an organization focused on the well-being of northern Indigenous women, they were interested in the research and its potential implications for community organizing, and they offered to house the project. We reached a reciprocal arrangement wherein I was given office space and access to their resources and, in exchange, helped with office activities over the three-month period. The NWA/NWT staff provided this research with invaluable support and expertise: they offered ongoing advice and insight as the fieldwork proceeded; they contacted their community networks to help recruit research participants; and they helped organize, design, and facilitate talking circles. The ongoing informal exchanges in the office as we discussed the research brought a rich reflexivity to the project that informed the evolving research design and analysis of the data.

I designed and organized interviews collaboratively and conducted interviews in a semi-structured way so that research participants would have the space to speak of their experiences in the way that they wished. I made interview transcripts available to participants so that they could retain a copy for their own records, and to give them an opportunity to amend or omit anything they had shared in the interview. Research participants were recruited through community radio ads and interviews, posters in the four communities, postings in regional email listservs, community organization networks, and the snowball method (i.e., some research participants recruited other research participants). I conducted thirty-three open-ended interviews, two talking circles, and one focus group, with a total of fifty-eight research participants. For the open-ended interviews, I identified four participant groups. These groups were adult women living in Behchokò, N'dilo, Dettah, or Yellowknife who self-identified as having a direct relationship to the diamond mines (target: 15–20 participants; actual: 20 participants), community workers (target: 6–8; actual: 8), diamond industry management (target: 2–4; actual: 2), and territorial government employees working in social and economic policy (target: 6–8 interviews; actual: 4).[12] Six interviews were held in Behchokò, the remainder in Yellowknife (see Appendix A for a list of interviews, talking circles, and the focus group).

I approached the first group, women who self-identified as having a direct relationship to the diamond mines, the majority of whom were Indigenous, as "primary informants." I defined "direct relationship" in the following ways: (1) women who have worked in the diamond mines; (2) women who have participated in a diamond-industry training program; (3) women who have a family member who has worked in the diamond mines; and (4) women who self-identify as having a direct relationship with the diamond mines. Most of the participants in this group had histories that spanned these criteria. For example, nine of these twenty participants initially identified themselves as partners/spouses of mine workers; in the course of the interviews, it came out that four of these nine had also worked in the mines themselves. Similarly, those six participants who initially identified themselves as mine workers often also shared stories about family members who worked in the mines. The multiple identities and relationships to the diamond mines inhabited by the participants crossed all four participant groups; I discuss this, and its implications for analysis, below.

These interviews were complemented by the two open talking circles hosted by the NWA. These talking circles were advertised on community radio stations, including Dene radio stations and the local CBC broadcast, through posters put up in the four communities, and through organizational networks and listservs. Between fifteen and twenty people attended the first talking circle, with ten of them speaking; between ten and fifteen attended the second talking circle, with eight of them speaking. Most of the participants in the two talking circles were Indigenous women, but participants included non-Indigenous women, as well as Indigenous and non-Indigenous men. As with the interviews, participant relationships to the diamond mines usually spanned the criteria outlined above, and participants drew upon the breadth of their experiences to reflect on the posted questions. The number of attendees is approximate because the talking circle was open to the community, so some participants came late or left early, some people came and chose not to speak, and others chose to speak but did not submit a consent form. Only those attendees who submitted completed consent forms are quoted.

Because some of the participants in the talking circle were also community workers, only one focus group was arranged solely for community workers, the second target group. The focus group, with ten participants, was designed as a two-hour guided group discussion about the community development and anti-violence landscape in Yellowknife; research participants were invited through Yellowknife community organization networks and coalitions focused on gender violence and community development. The focus group and eight community worker interviews were valuable sources of information in their own distinct ways. While the interviews provided the confidentiality and time for in-depth discussion, the focus group offered the opportunity for knowledge creation and transmission through group interaction (Kitzinger 1994). The third and fourth target groups were diamond-industry management and government personnel in positions related to diamond mining, respectively. The purpose of these interviews was to discuss company and government approaches to community impact, gender, and Indigenous relations to diamond mining. Interviews with diamond industry personnel focused on the two primary novel characteristics of the diamond-mining regime outlined above: that is, the FIFO model, and the regime's approach to consultation with and participation of Indigenous communities.

Interviews, talking circles, and the focus group provided concrete examples of the shifting social relations of subsistence, social reproduction, and capitalist production. Meanwhile, I conducted documentary analysis to develop an understanding of the political and economic context of these social processes. In Yellowknife, I collected print materials from the GNWT, the NWT Federation of Labour, the NWT and Nunavut Chamber of Mines, the NWT Status of Women Council, and community researchers. I also accessed documents online through federal and territorial websites, industry websites, and an online industry archive (NWT and Nunavut Chamber of Mines). These materials included policy, planning, and monitoring and evaluation documents (GNWT, NWT Federation of Labour, NWT and Nunavut Chamber of Mines, NWT Status of Women Council), labour force data (GNWT, NWT Federation of Labour; NWT and Nunavut Chamber of Mines), and socio-economic data (GNWT, NWT and Nunavut Chamber of Mines). These materials provided multi-scalar insight into the history of the subsistence economy, the territorial economy, the diamond industry, the move to FIFO, the diamond industry's role in the national and territorial economy, and regional strategies for employment and development.

The research participant sample operates at a different level of analysis than territorial data. The insights that emerge from interviews and focus groups do not lend themselves to generalizable conclusions for a number of reasons. First, the sample size is limited, and any impulses to quantify responses should be tempered by the multiple lenses through which research participants articulated their experiences. Indeed, in both interviews and focus groups, many of the research participants fit into more than one target research group; for example, some community workers also self-identified as women impacted by the mines, and some research participants' professional lives had included a combination of government work, work in diamond management, and/or community work. The numbers of research participants in each category,

listed above, reflect only the primary identity around which most responses centred and do not articulate the multiple identities through
which research participants engaged with the interviews. Thus, in
order to build analyses from these multiple identities rather than obscure them – given the small population of the region and the extreme
proximity (literal and figurative) of the diamond mines – in analysis,
I approached participant offerings as ethnographic narratives through
inductive thematic coding, not as quantifiable data. Furthermore, the
research participant sample was shaped through my own existing connections in the NWT and my collaboration with the NWA. While recruitment was also open to the general population of the region and
advertised through posters and radio, undoubtedly participation was
skewed toward those with an interest in gender justice and decolonizing community development, as well as those active in Indigenous and
community organizing. Thus, in providing a snapshot of forms of labour that often slip through the cracks of traditional labour surveys, the
interviews, talking circles, and focus group offer insight into the gendered gaps in many analyses of the mixed economy. They are grounded
sources of insight that enliven and complicate the FPE analysis in this
book, and exemplify the links between structural and embodied violence, as well as acts of decolonizing resistance.

V. Chapter Overview

The mixed economy in and around Yellowknife is a dynamic set of social relations that is reproduced through de/colonizing struggles. In
this book, I argue that the diamond-mining regime is a new imposition
on the daily and intergenerational social reproduction performed by
Indigenous women, an imposition that is sometimes violent and that
is met with resistance. The first part of the book lays the theoretical
(chapter 2) and historical (chapter 3) grounding for this argument. Part
II explores the contours of diamond mining in the global political economy (chapter 4) and in its relation to the northern mixed economy and
Indigenous governance regimes (chapter 5). This second part will help
readers understand the global context and the policy, economic, and
cultural infrastructure of the diamond-mining regime. It is in Part III
that the reader will find an empirical analysis of the gendered impact
of the diamond mines. The three chapters of Part III take up field research data to analyse the shifting social relations in the area around
Yellowknife through the development of the FIFO diamond mines.
Chapter 6 focuses on capitalist production; chapter 7, on day-to-day
social reproduction; chapter 8, on subsistence and intergenerational

community social reproduction. The conclusion draws together these analytical threads to offer a more systematic discussion of the restructuring of departments of production, and the relationship between restructuring social relations and violence against Indigenous women in the Yellowknife region. Here, I offer in greater depth the contributions of each chapter.

In chapter 2, "An Expanded Approach to Production," I develop this book's conceptual framework. The analysis draws upon Marxist categories of production but critiques these same categories for their exclusion and naturalization of social reproduction and subsistence. I argue that analyses of the social relations in the mixed economy in and around Yellowknife – and shifts in these relations – must incorporate both social reproduction and subsistence production at the same level of analysis as capitalist production. In so doing, I elevate social reproduction as a site of de/colonizing tension. I suggest that when analyses of production are expanded to include social reproduction and subsistence, social reproduction emerges as a locus of struggle between the reproduction of labour processes and ideologies oriented toward subsistence and those oriented toward the demands of capital. In the shifting relations between these departments of production, the gendered and racialized colonial violence of reorganizing for the pursuit of capital accumulation becomes evident. I present this expanded approach to production in the context of settler colonialism and the violent processes of racialization of Indigenous peoples in the north and in Canada more generally, paying particular attention to the gendered contours of these processes.

In chapter 3, "Wıìlıìdeh's Mixed Economy," I take up the categories outlined in chapter 2 and engage them in an analysis of the development of the mixed economy in and around Yellowknife from the mid-twentieth century to today. In taking an FPE approach, I emphasize the gendered and racialized (and, arguably, violent) nature of the shifts in departments of production through the recent history of the mixed economy. These shifts have occurred in part because of the gold mines established in Yellowknife, but more so as a result of the activist role the Canadian state has taken in restructuring northern Indigenous social reproduction. I argue that in the Dene's traditional and early mixed economy in the area around Yellowknife, subsistence production and social reproduction were one and the same. Largely through Canadian state activities, social reproduction was feminized and isolated from subsistence, which was, for its part, racialized and restructured. The Canadian state played an activist role in restructuring Dene people's social reproduction toward the demands of capital, but this restructuring was not matched with sustained efforts to integrate them into capitalist production (that is, until the diamond mines were established). When these processes are analysed through an expanded conception of production, it becomes clear that Dene social reproduction was targeted as a site of struggle between subsistence and capitalist production and that contemporary feminized responsibility for social reproduction (to the extent to which it has been internalized) is the result of a relatively recent rupture in social relations in the area around Yellowknife.

In chapter 4, "The Global Political Economy of Canadian Diamonds," I locate northern Canadian diamond mines within the historical and contemporary global ambit of diamonds, those enigmatic gems. Diamonds, and the social relations of the mines, markets, and distribution networks that bring them from kimberlite deposits to romantic tableauxs around the world, are both exceptional and violently mundane in their continuities with processes of extractive dispossession across space and time. I trace the material and cultural relations that came to conflate diamonds with romantic love, beginning with the discovery of the Eureka Diamond in South Africa and the building of the De Beers empire. Moving to the contemporary, I trouble the category of "conflict" mining, which relies on racialized tropes to characterize violence as an aberration from "normal" extraction. It was in the context of the crisis of "conflict" mining in the late 1990s that the Canadian diamond industry developed as a responsible alternative to the stained reputation of diamonds mined in the Global South. Chapter 5, "The NWT Diamond-Mining Regime," takes a close look at the so-called responsibility of the Canadian diamond alternative; it explores the extent to which diamond mining represents a "new extractivism" by examining the dialectic between the diamond mines and northern Indigenous movements and the resulting multi-scalar local governance regimes. I compare the neoliberal diamond-mining regime with the gold mines that dominated Yellowknife's political economy for most of the twentieth century and that abruptly shut down just a few years before diamond mining began. Compared to the gold mining regime, which did not consult with Indigenous people and operated through a model of ostensibly "separate development," the diamond-mining regime represents a new approach to extractive/Indigenous relations, articulated through its institutional arrangements – in particular, socio-economic agreements (SEAs) between the GNWT and impact benefit agreements (IBAs) between Indigenous communities and the diamond-mining companies. While SEAs and IBAs are positioned as a means of resolving the tensions between extractive production and local Indigenous communities, I argue that the individualized, masculinized approach to Indigenous engagement expressed in these agreements – a reflection of the diamond-mining regime's general approach to Indigenous relations – obscures the impact of the diamond mines on the social relations of subsistence and social reproduction.

It is this impact that I explore in the third part of the book, which draws on the insights of research participants to examine northern Indigenous women's experiences with the diamond mines. Chapter 6, "Time, Place, and the Diamond-Mining Regime," examines labour in the diamond mines and the ways in which those mines interact with the capitalist labour market in the NWT's mixed economy. The FIFO diamond-mining regime is, in some ways, in continuity with the history of extraction in the region, but it also represents a new – and distinctly gendered and racialized – approach to capitalist production because of its intensified flexible approach to labour and because of how it spatially articulates capital's separation of social reproduction from capitalist production. I combine an analysis of the diamond mines' operational structures with an analysis of women's experiences of working at those mines as a way of illustrating the sometimes violent gendered and racial hierarchies of labour that straddle sites of extraction and sites of "home" as they have been restructured through the FIFO diamond-mining regime.

In chapter 7, "Social Reproduction and the Diamond Mining Regime," I argue that the FIFO masculinist approach to capitalist production pursued by the diamond-mining regime has resulted in a reorganizing of social reproduction as performed by the region's Indigenous women. Drawing on research participants' contributions, I outline two manifestations of the reorganization of relations of social reproduction for the pursuit of capital accumulation in the form of FIFO diamond mining. I identify a (contested and uneven) reorientation of social reproduction toward the demands of capital and away from subsistence. This reorientation is accompanied by a restructuring of social reproduction through the spatial articulation – in the form of FIFO extraction – of the ideological divide between social reproduction and capitalist production. In the context of the regional mixed economy, a central result of this restructuring has been to "make nuclear" social reproduction – that is, to intensify and localize household-level social reproduction at the expense of the inter-household linkages that support the reproduction of community social relations. However, this restructuring has occurred incompletely and in relation to the everyday ways in which Indigenous women reproduce and prioritize inter-household linkages. Those ways constitute acts of resistance and are discussed in this chapter and the next.

I turn to subsistence and community-level intergenerational social reproduction in chapter 8, "Diamonds, Subsistence, and Resistance." There I explore the relationship between the FIFO diamond-mining regime and subsistence production, examining the impact the extractive regime has had on land-based social relations. I argue that while FIFO structures manage the tension between the demands for mobility by contemporary capital and the material fixity of resource extraction, the greater time/place tension at play in the mixed economy is between the capitalist imperative to annex space with time, on the one hand, and the place-based social relations of subsistence, on the other. I explore the reorientation, or increased orientation, of Indigenous women's labour toward the demands of capital as a result of the imperatives of the diamond-mining regime, tracing restructuring and resistance along three lines of inquiry: sharing, community, and kin; subsistence and education; and the land and the body. I contend that this reorientation is violent on both structural and embodied levels; however, this shift is also incomplete, contested, and shaped through resistance. The day-to-day labours that socially reproduce and strengthen a subsistence orientation in the face of the totalizing impulses of capital and the Canadian state exemplify the ways in which the constrained – and sometimes messy – choices made at the level of social reproduction shape the social relations that make up a given regional political economy and, in the context of this case study, reproduce the regional mixed economy.

A theme running through this book, particularly the final two chapters, concerns the relationship between the restructuring of northern Indigenous women's labours and structural and embodied violence. In the Conclusion, I take up this relationship, asking what of this analysis might be applicable more broadly. The FIFO model encourages an intensified hierarchical ordering of masculinized capitalist production over feminized social reproduction. This restructuring is tied to the exploitation of land, which manifests itself as structural and embodied violence that is simultaneously racialized. Experiences of violence by Indigenous women in Canada demonstrate that the structural and the embodied are not discrete and are often experienced at the same time. I use this chapter to engage in broader conversations about violence against Indigenous women in Canada, as well as the gender and racial violence of extraction. My hope is to reveal the insights and the emerging questions associated with this study, while looking ahead to potential future avenues of inquiry, collaboration, and social change.

PART ONE

Theorizing the Northern Mixed Economy

2 An Expanded Approach to Production

We need to value young women in the same way that we value diamonds. They're not disposable. And yet we're still treating them [that way]. And I felt this as a young person: a resource, but not a separate agent.

– Interview 104

Introduction

In this chapter, I outline my framework for theorizing the shifts in the Wıìlıìdeh mixed economy as a result of the diamond mines. Much political economy limits its frame of reference to capitalist labour, leaving aside all labour that sits outside or alongside the boundary of that which is capitalist. The violence of the pursuit of capital accumulation – in particular, its racialized and gendered contours – is obscured by such a narrow concept of labour. The framework proposed in this chapter is, alternatively, an expanded conception of production that approaches social reproduction, subsistence production, and capitalist production as interconnected forms of labour at the same level of analysis. As I apply this framework in the chapters that follow, social reproduction will emerge as a site of tension between subsistence and capitalist production. It is there, and in the relations that weave through these multiple labours, that one can trace the gender violence that ensues when labour power is reorganized in the pursuit of capital accumulation. Colonial violence is met with decolonizing resistance.

The first part of this chapter introduces an expanded approach to production, for which I draw upon Indigenous and anti-racist feminisms as well as feminist political economy (FPE), which has grown out of feminist historical materialism (Young 1980; Jenson 1986; Vosko 2002). In analysing the contemporary shifting relations between departments

of production under the FIFO diamond-mining regime in the Yellowknife region, I take an expanded approach to production (Seccombe
1992), an analytic framework that approaches value-producing labour
(capitalist, or productive, labour) at the same level of analysis as social
reproduction. FPE, anti-racist, and critical theorists have long called for
and contributed to theories that integrate social reproduction into the
"political" and the "economic." However, for a fulsome approach to
social relations in the region of study at hand, subsistence must also
be included as an element of the mode of production. Subsistence,
recall, encompasses land-based activities oriented toward the daily
and intergenerational well-being of the collective, a concept I elucidate
upon below. When subsistence is introduced at the same level of analysis as social reproduction and capitalist production, political economic
analysis is unhinged from the axis of capitalist production, thus expos-
ing the *ongoing* shifts in power relations between different labours. The
historical analysis in this book will demonstrate that in the traditional
economy, social reproduction and subsistence are indistinguishable
in that both are forms of labour directed toward the social needs of
the community. Social reproduction *becomes* distinct from subsistence
through Canadian state activities that, on the one hand, racialize and
marginalize subsistence, and on the other, feminize and restructure
social reproduction toward a Western capitalist model. Thus, contemporaneously, social reproduction sits as a site of tension between subsistence and capitalist production. It is through this analytical lens that
we are able to understand the material impetus behind the Canadian
state's interventions into Indigenous women's activities as they relate
to social reproduction in the Yellowknife region.

In the second part of this chapter, I move from a discussion of *what*
(specifically, what counts as production) to a discussion of *how* – that is,
how the social relations of reproduction and production are approached
through settler and Indigenous social relations in the northern mixed
economy. I draw upon Indigenous theories of subjectivity to ground my
analysis of Indigenous/settler relations, taking a place- and practice-
based approach to what it means to be Indigenous, in opposition to
state-sanctioned, static approaches to Indigenous identity. Indigenous
identities are constructed through decolonizing relations and practices;
but simultaneously, they have also been formed through the violent racialization processes of past and present-day settler colonialism. I attend
to the continuities of settler colonial dispossession that have shaped the
northern political economy; at the same time, I disrupt the presumed lin-
earity of those processes by turning to Tłı̨chǫ Dene relational approaches
to history that foreground place-based relations rather than time-based

development, and make space for honouring the multiple sites of decolonizing resistance that reproduce the northern mixed economy.

I. An Expanded Conception of Production

The concepts of production, labour, and work may seem straightfor-
ward. However, throughout the history of capitalism, the ways in
which certain forms of labour – particularly gendered and racialized
labour – are obscured (i.e., do not count as work in a particular society) have been common means of exploitation. In taking an expansive
approach to these terms, I am able to honour the feminized and Indigenous labours so often unaccounted for in measures of northern production even while tracing the settler colonial tensions at play within
and between different forms of labour. Here, production refers to the
combination of forces and social relations required to create a social
good, while labour refers to the human effort involved in those processes. But of the myriad activities that capture human energies on a
daily basis, which ones actually count as labour? Academics and poli-
cy-makers (among others) tend to define labour as work for a wage – a
historical tendency so strong that any other way of approaching labour
seems jarring. For example, the Canadian Labour Force Survey equates
labour with employment, which is defined as "work for pay or profit"
(Government of Canada 2018b). The International Labour Organization uses a similar definition, though it extends the concept to "work
carried out in exchange for remuneration payable in cash *or in kind*"
(ILO 2018, my emphasis), which is a nod, at least, to the diversity of
economic exchange around the globe today.

Critical political economy, rooted in Marx's analysis of labour under
capital, offers a more expansive and substantive definition of labour. For
Marx, labour, in the abstract, is a universal condition of human existence –
indeed, it is nothing short of the expression of human essence, which
manifests itself through the expenditure of energies in mediating the
"metabolism between man and nature" (Marx 1976: 133). However, subsequent Marxian analysis has narrowed its subject through a tendency
to make *capitalist* labour the sole source of inquiry. Capitalist labour, for
Marx, refers specifically to labour that is exploited for the creation of surplus-value, wherein "the worker produces not for himself, but for capital" (644).[1] Labour that is not altogether capitalist – that is, labour that is
not performed directly for the accumulation of surplus-value – is often
invisibilized, naturalized, or assumed to be on the brink of extinction, if
not already extinct. This omission obscures a thriving, varied, and unruly
set of social relations that challenge capital's hegemony, from day-to-day

labours that resist the pressures of exploitation, commodification, and privatization to organized and well-established alternatives to Western capitalism, both of which we see in the northern mixed economy.

To uncover this varied set of labours (which are, while not outside of capitalism, structured and performed in ambivalent relation to capitalism), let us turn first to Marx's approach to the reproduction of labour-power (what I refer to as social reproduction). In *Capital*, volume 1 (1976), Marx writes: "The maintenance and reproduction of the working class remains a necessary condition for the reproduction of capital. But the capitalist may safely leave this to the worker's drive for self-preservation and propagation ... From the standpoint of society, then, the working class, even when it stands outside the direct labour process, is just as much an appendage of capital as the lifeless instruments of labour are" (718–19).

Marx's assertion encompasses both the strengths and the weaknesses of his analysis of the reproduction of labour-power. Marx conceptualizes the production of capitalist value[2] and the reproduction of workers (specifically, workers' labour-power) as indivisible elements of the same process: the valuation of capital. He contends that capitalist production depends on the reproduction of labour-power that occurs outside of the workplace (Ferguson and McNally 2013: xl). This points to the limited analytical capacity of a framework that privileges "real" (value-producing) wage work over social reproduction performed informally in communities and households.

However, while Marx acknowledges capitalism's dependence on social reproduction, he also naturalizes its constitution. For Marx, structures of social reproduction are formed through the social relations onto which capitalism is grafted (i.e., the Western male-dominated nuclear family) and the imposition of the demands of capital. Marx naturalizes the former and assumes that the latter is the historical agent; social reproduction is thereby left depoliticized, inert. Indeed, for Marx, if there is capitalist production, there will be capitalist social reproduction (1976: 711). Marx's – and subsequent Marxian – naturalization of social reproduction leaves an analytical gap: *social* reproduction is *de*socialized by being characterized as biological and natural and thus devoid of power relations and struggle. Indeed, Marx writes, "the organization of the capitalist process of production, once it is fully developed, breaks down all resistance ... In the ordinary run of things, the worker can be left to the 'natural laws of production'" (899). Thus, Marx assumes that social relations in capitalist spaces will succumb to the logic of capital. This assertion denies social reproduction as a site of struggle; equally, it denies non-capitalist relations of production, such as subsistence, that persist both inside and outside of capitalist (or mixed) spaces.[3] The assumption that fully developed capitalist production "breaks down all resistance" posits a teleological binary between that which is capitalist and that which is not, as if a combination of capitalist and non-capitalist production is a *temporary* phenomenon in the process of capitalist colonization. This binary, however, cannot account for the mixed economies that have sustained themselves even within the ambit of global neoliberal capitalism. Indeed, in the Yellowknife region, the mixed economy offers relative stability through the boom and bust (i.e., ruptures) of various resource extraction projects. The strength of a mixed economy should not be confused, however, with its invariability or inevitability. Rather, a mixed economy is a dynamic site of struggle wherein the day-to-day social reproduction of community and land-based relations refuses capitalism's tendency to "break down all resistance" by protecting the modes of life that make the mixed economy "mixed."

Thus, I approach both subsistence production and social reproduction at the same level of analysis as capitalist production. Drawing upon Indigenous and feminist theories of subsistence (Nahanni 1977, Abele and Stasiulis 1989, Adams 1995, Mies and Bennholdt-Thomsen 1999, Kuokkanen 2008, Coulthard 2010, Kuokkanen 2011), as well as critical empirical studies of subsistence in the Circumpolar North (Asch 1977, Watkins ed., 1977, Usher et al. 2003, Abele 2009a, Harnum et al. 2014), subsistence, in this text, refers to land-based production as it is performed by Dene, Métis, and Inuit peoples in the NWT. Subsistence is about more than just securing daily maintenance (as opposed to the capitalist accumulation of surplus-value); it also involves an orientation toward collective, not individual, need rooted in an ethic of relationality to and care of the land and non-human animals. These labours are grounded in place-based histories and sets of meaning – in a "grounded normativity" (Coulthard and Simpson 2016). As will become clear as I elucidate upon the elements of this framework and their relationships, to characterize subsistence as non-capitalist is not to deny the local workings of global capitalism, but to identify and honour labours that are not driven by or structured primarily in terms of the accumulation of surplus-value.

When subsistence is introduced at the same level of analysis as social reproduction and capitalist production, political economic analysis is unhinged from the axis of capitalist production, thereby making visible the ongoing de/colonizing shifts in power relations between departments of production in the mixed economy. To attend to the shifting social relations of the mixed economy as they are shaped by the imperatives of social reproduction, production for subsistence, and the accumulation of surplus-value, I present the framework in Figure 1.

Figure 1. An expanded conception of production.

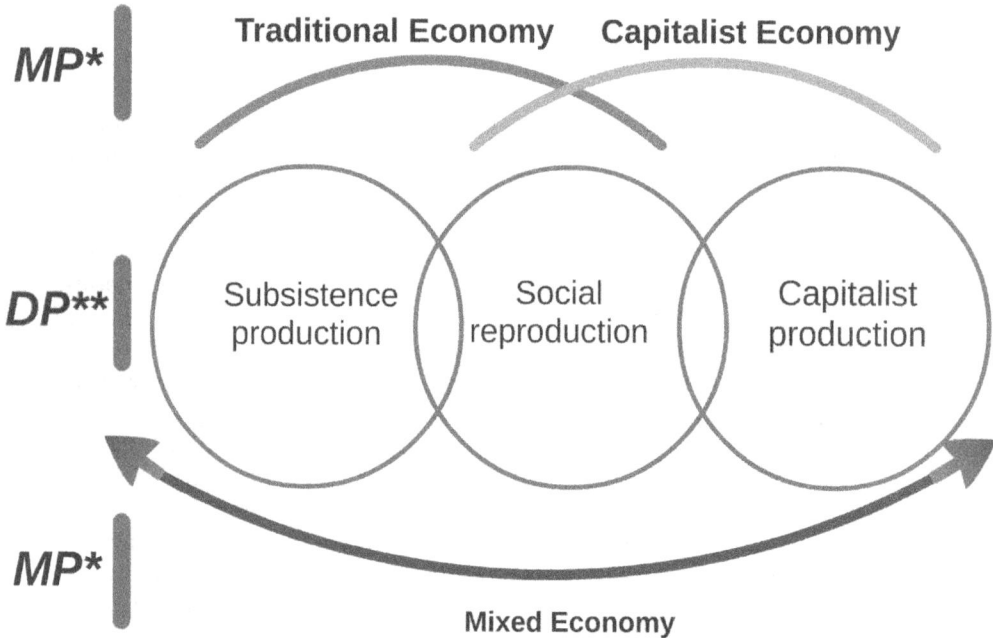

*Mode of Production
**Departments of Production

This framework owes its theory to Indigenous, FPE, and anti-racist scholarships that have long advocated for and contributed to theories that integrate social reproduction and subsistence with the "political" and the "economic." In developing this approach I have followed Coulthard's (2014) suggestion that mode of production be conceived broadly as a "mode of life." He argues that mode of production "accurately reflects what constituted culture in the sense that the Dene deployed the term, and which our claims for cultural recognition sought to secure through the negotiation of a land claim" (65). In this formulation, mode of production is tied to both ontology and place, thus making space for the ways in which the labours of the mixed economy are rooted in its land. Labours of the mixed economy transcend Western labour categories and, as Abele (2006) notes, can be at once challenging and nourishing, necessary and recreative.

The framework owes its form to Wally Seccombe (1992), who, alongside many notable feminist political economists (Jenson 1986, Mies 1986,

Vosko 2000, 2002, Federici 2004), suggests that rather than approaching the production of labour-power (a component of social reproduction) as a sphere in isolation from capitalist production, or, alternately, characterizing social reproduction as capitalist production itself, one should embrace a broad conception of the socio-economic. By approaching social reproduction as a *distinct* form of labour with the same ontological status as capitalist production, Seccombe (1992) offers an approach to the relations between different forms of labour that does not tacitly subordinate one to another or attempt to claim one as the other.[4] This opens theoretical space for attention to the unique character of social reproduction and its ambivalent relationship to capitalist production, for social reproduction is both *necessary to* capital relations and *outside of* capital relations (Vosko 2002). Seccombe's framework, rooted in an analysis of Western Europe, does not include subsistence; however, I extend his "distinct but of the same ontological status" approach to production for subsistence in the NWT mixed economy.

I elucidate on this theoretical framework at two levels of analysis, illustrated in the above figure: mode of production, and department of production. In Marxian political economy, mode of production refers to the combination of forces of production and relations of production organized to sustain and reproduce a "distinctive mode of appropriating surplus labour" (Jessop 1982. Following Coulthard (2014) and Bannerji (2005), I am interested in the place, cultures, and people that make up a given mode of appropriating surplus labour, and thereby approach a mode of production as a "mode of life" (2014). Thus, *contra* economistic interpretations of this definition, I emphasize that relations of production are *social* relations and must be approached through the ideological and cultural structures through which they dialectically operate (Bannerji 2005). Furthermore, as Jairus Banaji (1977) brings to light, Marx has two distinct uses of the term "mode of production": one to refer to specific and concrete "technical processes of production," and the other to refer to global stages, or epochs, of social development (5). While I am attentive to the relationship between the regional political economy under study and the global epoch throughout this book, my usage of the term "mode of production" within this framework falls squarely in line with Marx's first meaning. I have developed this framework to illustrate the shifting social relations in the NWT mixed economy. In a more general analysis of the Canadian – or, indeed, global – political economy, one would position the mixed economy as a regional socio-economic formation operating *under* capitalism. However, the analytical framework illustrates the ways in which the traditional northern Indigenous economy and the capitalist economy come together in the contemporary mixed economy in the Yellowknife region. As will become clear below, the traditional economy depicted in Figure 1 no longer exists as such, but rather is included in the diagram as a comparative and explanatory set of social relations used for analysing the genesis of the mixed economy. I first discuss modes of production and then turn to departments of production.

a) Modes of Production

In Figure 1, the semi-circles framing the circles (which are departments of production) represent three modes of production: the traditional economy, the capitalist economy, and the mixed economy. The traditional economy is comprised of subsistence production and social reproduction. The capitalist economy – defined as an economy oriented toward the creation of surplus-value through the exploitation of labour-power (Marx 1976) – is comprised of capitalist production and social reproduction. While I separate the traditional economy and the capitalist economy for the purpose of clarifying analytical categories, I am not arguing that the capitalist economy developed separately from the traditional economy; rather, the capitalist economy developed through initial and ongoing exploitation and appropriation of traditional and mixed economies. The mixed economy combines the capitalist economy and the traditional economy. The mixed economy column is bordered by arrows to denote its dynamism: the historical and contemporary shifts in balance between capitalist production and subsistence production. For example, during the fur trade, subsistence dominated the northern mixed economy, whereas contemporary resource extraction has shifted social relations toward capitalist production.

I use the term "traditional economy" as this is the term used in local Dene (Tłı̨chǫ and Wıı̀lıı̀deh) publications (see YK Dene First Nation Advisory Council 1997; Zoe 2005). The term "traditional" is not in reference to a static, unchanging concept of a socio-economy; rather, it refers to socio-economies imbued with place-based history and meaning, a concept that straddles Western political economic distinctions between the material and the ideological. Traditional, or land-based, economies vary around the globe, shaped as they are by the land and resources with which they operate and by the histories of cultures of their place. In Canada, the traditional economy generally refers to Indigenous socio-economic formations that are non-capitalist and based on the relationship between people, animals, and the land, economies that vary widely from the coastal regions of the Atlantic and Pacific to the Great Lakes, the plains, and the Subarctic and Arctic. Asch (1977) characterizes the northern Indigenous economy in the NWT before the fur trade as a "to-tal economy both in the sense of production and circulation of goods" (49). The rocky land and cold climate meant that traditional economies relied primarily on hunting, trapping, fishing, and foraging; practices of drying and preserving meats and making teas and medicines from the local bush; and the use of skins and furs for clothing and household goods. George Blondin (1997) describes how, in the Chipewyan Dene traditional economy, these practices were deeply relational:

> We are people of the land; we see ourselves as no different from the trees, the caribou, and the raven, except we are more complicated. Long be-fore Europeans arrived in our land, The Dene lived all over the 724,000 square kilometres of Denendeh. The largest group was the Chipewyan; they hunted and fished in the Lake Athabasca area, in what is now north-ern Saskatchewan, northeastern Alberta, and farther north into the Barren Lands. They followed the migrating caribou and were called "Caribou Eaters" by the more southerly Cree. (18)

From the fur trade onward, the northern economy maintained its land-based orientation, but it was reproduced through a mixture of sub-sistence and the use of externally produced goods acquired through exchange: a mixed economy (Asch 1977). However, as argued by In-digenous elders and scholars (Watkins, ed., 1977; YK Dene First Nation Advisory Council 1997), because the fur trade did not substantially dis-rupt Indigenous socio-economies, the pre–Second World War northern economy can still be understood as one oriented toward subsistence (DiFrancesco 2000; Gibson 2008).

The northern mixed economy developed through a series of twen-tieth-century extractive projects that introduced capitalist relations of production. The development of a mixed economy in the North is usually defined temporally by scholars as postwar. Significantly, im-mediately after the Second World War, Indigenous communities that had been predominantly nomadic moved into government social hous-ing in sedentary communities. However, "although this move brought deep and important changes to northern life, it did not lead to an end of life on the land. Instead, the way of life that Indigenous people had followed for centuries was adapted to reflect their new constraints and opportunities" (Abele 2006: 185). Today, on-the-land activities are still pursued by Dene, Métis, and Inuit living in the bigger cities of the NWT, like Yellowknife; that said, it is especially the small communities of the NWT that rely on subsistence for a large proportion of their so-cial needs.[5] Protecting, pursuing, and respecting the mixed economy is an important and ongoing decolonizing practice in the NWT. While relationships to capitalist production within the mixed economy vary across time and place, it is the commitment to pursuing subsistence activities and making them visible that characterizes contemporary northern life.

b) Departments of Production

These three modes of production – the capitalist economy, the traditional economy, and the mixed economy – are populated by three departments of production: social reproduction, subsistence production, and capitalist production. Figure 1 depicts these three departments as overlapping as a means to represent the permeability of the boundaries between these categories (e.g., child care performed as wage labour, or hunting a caribou that is subsequently directly consumed by a care unit, shared with friends and community, and sold). Indeed, throughout this book I will be using the term "department" not to depict discrete labour relations but rather as a heuristic device to help identify the different imperatives and social relations that shape people's labours in the mixed economy – labours that often overlap. Social reproduction sits as a category of labour necessary for both subsistence and capitalist production; as a site of de/colonizing struggle, it is a space of both colonial violence and creative decolonizing resistance.

I use the term "capitalist production" to denote the social relations of "productive" labour in the Marxian sense – that is, labour that is exploited for the creation of surplus-value, so that "the worker produces not for himself, but for capital" (Marx 1976: 644). Marx explains that productive labour has nothing to do with its own utility; instead, "labour is only productive so long as it is producing its antithesis" (1971: 79). This definition has been used to marginalize women, as well as anyone not engaged in "productive" or paid work, when political engagement with work is being analysed. FPE theorists have challenged this exclusion as both faulty and oppressive, and I agree with them. However, I suggest that the problem is not with the definition of productive, or capitalist, labour; rather, it is that the broad term *labour* is often reduced to these same parameters. Thus, I use the concept *capitalist production* here not to limit analysis but rather to do the opposite: to distinguish *different* relationships between labour and capital and to avoid privileging productive labour over other forms of labour.

A theory of social reproduction is one that acknowledges the relationship between production for surplus and the reproduction of daily life (Vosko 2002; Bhattacharya 2017). However, the term, "social reproduction" is not deployed consistently in FPE literature (as noted by Vosko 2002; Bezanson and Luxton 2006; and Ferguson 2008). Beyond the expected and often fruitful divergences and debates among theorists, there is an analytical cloudiness that hampers the transformative potential of this concept. Meg Luxton (2006) writes that "by itself, social reproduction offers little more than a fancy term to describe the ordinary activities of daily life. Too often, conventional feminist use of social reproduction still focuses on women's work in the home, leaving vague its relationship to the complementary work (often done by women for pay) provided by state services such as education and health care or in the market" (6).

When the term "social reproduction" slips into shorthand for women's work in the home, it loses the conceptual clarity required to challenge the erroneous separation of social reproduction from capitalist production. Furthermore, assuming that social reproduction refers to feminized unpaid labour in one's own home privileges a Western nuclear conception of the household, thus obscuring the social reproduction – and the transnational racialized and colonial structures through which it operates – that is done for pay and not for pay through inter-household kin and community networks that transcend the picket fence of the imagined family home.

How, then, within an expanded conception of production, does one articulate a concept of social reproduction that takes as its starting point the indivisibility of departments of production – as parts of a whole – but also acts as a precise category of labour through which to explore the shifting racialized and gendered relations between departments of production in a given place and time? The first step in addressing this issue is ascertaining whether the term "social reproduction" refers to the social reproduction of the mode of production in its totality (thinking here of a "mode of production" as epochal (Banaji 1977)) or whether social reproduction refers specifically to the work done to reproduce labour-power, with the cognizance of its dialectical relationship with value-producing labour, and subsistence.

In an example of the former conception, Ferguson characterizes social reproduction as "the concept of an expanded mode of production, whose essential unity lies in a broad definition of labour. That definition … incorporates both the value-producing labour associated with the waged economy and the domestic labour (typically performed by women) required to give birth to, feed and raise the current generation of workers and the children who will comprise the future workers" (2008: 44). Bezanson and Luxton's (2006) definition of social reproduction is narrower: "The concept of social reproduction refers to the processes involved in maintaining and reproducing people, specifically the labouring population, and their labour power on a daily and generational basis. It involves the provision of food, clothing, shelter, basic safety, and health care, along with the development and transmission of knowledge, social values, and cultural practices and the construction of individual and collective identities" (3). It is notable that even within the narrower definition offered by Bezanson and Luxton, social reproduction is not synonymous with biological reproduction or nuclear forms of domestic labour. It includes educative and cultural labours, which, alongside caring labours, are certainly performed at multiple scales, from institutions (such as schools, art centres, and community halls) to informal community networks and inter- and intra-household linkages. Defining social reproduction specifically as the reproduction of the labouring population rather than as the reproduction of the mode of production in totality is a choice that need only be one of semantics. What is necessary is an analysis of the *relationship* between the two. Either formulation makes visible social reproduction and the gendered hierarchies of labour between and within departments of production. So I use the term "expanded conception of production" to refer to an analytical framework that encompasses capitalist production, social reproduction, and subsistence production, and I approach these different forms of labour relationally and at the same level of analysis. I use the term "expanded conception of production" here in the same way that Ferguson uses the term "social reproduction theory," that is, to speak to social relations of a given place and time in their totality. Social reproduction as a category of labour or department of production refers here to the narrower conception of paid and unpaid work performed for the purpose of daily and intergenerational reproduction of a given community or population. In the context of the Indigenous relations and ontologies of the northern mixed economy, then, social reproduction refers to the multi-scalar caring, biological, cultural, and land-based labours that reproduce the people and the social and land-based relations of northern communities. These persist in relation to, and are sometimes one and the same with, the capitalist and subsistence labours that together make up the expanded conception of production in the mixed economy.

The concept *social reproduction* was developed in response to scholarship privileging capitalist labour over non-value-producing labour. As such, the relationship between social reproduction and other forms of non-value-producing labour – for our purposes, subsistence – is undertheorized, due partly to tendencies toward "race-blind" and Eurocentric theorizing in political economic traditions. Thus, the third category of production, subsistence production, is perhaps the most crucial to delineate, though also the most difficult. Subsistence, in this book, refers to land-based labour organized through Indigenous institutions, ontologies, and communities. This is distinct from the way the term is used in political economy, wherein subsistence refers to the acquisition of the immediate needs required for a worker to reproduce her/himself. In both Indigenous scholarship and global studies of diverse economic formations, *subsistence production* denotes broad socio-economic relations and processes as well as a real and aspirational labour orientation (Mies and Bennholdt-Thomsen 1999). While subsistence is practised around the globe, I am approaching subsistence in a more specific regional and historical sense as labour most often performed by northern Dene, Métis, and Inuit and oriented toward the social needs of the household or community (recognizing that "community" is a flexible and shifting term), attached to places and the plants and animals of those places, and practised through local social relations. In chapter 3, I will detail the ideological and material specifics of subsistence in the mixed economy in and around Yellowknife. Here, I focus on its theoretical definition as it relates to the expanded conception of production framework.

Unlike capitalist production, subsistence is based on social need rather than individual accumulation. I use the term "social need" rather than simply "need" because subsistence, or the traditional economy, should not be equated with hand-to-mouth living. A socio-economy can be rich and abundant without being based on the accumulation of value, and need is not simply a physiological state; it is also a social one and can include various comforts, entertainment, art, and culture. Like social reproduction – which can be, at once, labouring and emotive, exhausting of labour-power and interpersonally enriching – subsistence is much more than a necessary activity for biological or economic sustenance. Abele (2006) writes that "'going on the land' is physically arduous and sometimes risky, but it is not typically understood as 'work.' Rather it is recognized as an activity that contributes a great deal to physical, emotional and mental well-being" (187). Social reproduction and subsistence are also both oriented toward providing for social need; while the two concepts come out of different traditions and have distinct relationships to capitalist production and locations within the mixed economy, there is no inherent distinction between the two. Indeed, when subsistence production constitutes a total economy (as it was in the traditional northern economy), the concepts collapse into each other. The distinction between subsistence production and social reproduction emerges imperfectly in the mixed economy, wherein – largely through state interventions – social reproduction is isolated from subsistence production as a feminized labour process oriented toward daily and intergenerational reproduction of (presumed capitalist) labour-power. Subsistence, for its part, is marginalized as racialized extra-economic activities that are outside the mode of production. It is in this incomplete process of separation – an ongoing site of contestation that is a focus of this book – that subsistence and social reproduction are made distinct and that social reproduction becomes a site of colonial capitalist struggle.

An expanded conception of production facilitates an analysis of the ways in which these different forms of labour interact and sometimes overlap, thereby bringing a corporeal intensity to the violence of restructuring local social relations for the pursuit of capital accumulation while also opening space for stories of resistance. Indeed, it is through the relational totality of these three departments of production that the gender and racial violence of reorganizing labour-power become evident, as well as resistance to this violence. Many participants in this research project expressed the different forms of resistance they practised, whether it was through orienting their labour toward subsistence, through the privileging of interpersonal relations over the demands of capital (in the form of diamond mining), or, as was often the case, through a combination of these tactics. I argue that labours oriented toward the social reproduction of Indigenous communities and toward subsistence constitute resistance to the totalizing pressures of capitalism as well as to the Canadian state, which protects and reproduces the mixed economy. They exemplify the ways in which Indigenous people in Canada resist colonization through daily enactments of work, relationships, community, and culture (Maracle 1988; Kino-nda-niimi Collective 2014). For example, in describing strategies for building relationships and staying healthy at the mines, Iris, an Inuk woman who had worked on and off with the diamond mines, described mine camp life in this way: "In the night-time, after that [housekeeping work], we go fishing in the lake. We go pick berries, take some heather, make some tea. Lots of heather … You gotta take the heather and burn the heather. And you have to take lots of heather because it burns really fast, but the smell is so good. And the taste, oh!" (Interview 101). For Iris, engaging in subsistence practices at the mine site was a source of interpersonal, emotional, and physiological well-being.

Doris, a Dene woman, drew on her expertise in subsistence production in a different way. Doris experienced sexual harassment at the mine and found the shift work incompatible with her responsibilities to her extended family. For these reasons, she drew on her expertise in sewing and beadwork to divorce herself from economic reliance on the mine:

> Working at the mine, if I wasn't working, if I was at home for two weeks, that was all I did. Just sew. Sewing saved me from a lot, you know. Before, before my husband had a full time job, I used to make money on the side, sewing non-stop. Sometimes I would be up twenty-four hours straight. Just sewing, sewing, sewing. So I could sell it and buy groceries. I did a lot of nights like that, just to buy groceries. So sewing saved us from a lot of hard times ... Because he wasn't very good with his paycheque ... So I would get back into sewing again just to make up the difference. So I love sewing. Some days, I just sit around and I can just imagine all these designs in my head. Floating by. (Interview 111)

For Doris, sewing and beadwork had been a significant source of income, stability, and peace throughout her life, including during her relationship with a man who worked at the mines and during her own time working at the mine.

Both Doris and Iris demonstrate that subsistence and social reproduction are far more dynamic, varied, and resilient than Marx's understanding of a sphere of labour dictated by biological forces and the needs of capital. In approaching social reproduction and subsistence at the same level of analysis as capitalist production, and theorizing these different forms of labour through the contradictory ideological and material relations that bind them, one can avoid relegating subsistence to a "cultural" silo and social reproduction to an "emotive" or "private" silo, thus obscuring important strategies of resistance that shape social relations in the region of study. Instead, one can look to the ways in which Indigenous women's labour processes sustain the minority ways of life that contribute to the mixed economy.

II. Indigenous Practices of Resistance

While this framework unhinges political economic analysis from the axis of capital, certainly, one must move forward recognizing that labour is far more than its location in a mode of production. Labour is qualitative, cultural, and subjective (Bannerji 2005), as well as imbued with meanings that transgress the capitalist/non-capitalist boundaries. This recognition follows the decolonizing imperative outlined by Coulthard (2014), who argues that "any strategy geared toward authentic decolonization must directly confront more than mere economic relations; it has to account for the multifarious ways in which capitalism, patriarchy, white supremacy, and the totalizing character of state power interact with one another to form the constellation of power relations that sustain colonial patterns of behaviour, structures, and relationships" (14).

Indigenous identity, as a racialized subjectivity, operates at a distinct level of analysis from subsistence, though subsistence, as it is approached in this study, is performed primarily by Indigenous peoples. Indigenous subjectivities and subsistence are tied together through histories and ontologies grounded in the land; but, grounded though they may be, subsistence and Indigenous identity are dynamic categories that shift over time. Coulthard (2010) and Alfred and Corntassel (2005) approach the category of "Indigenous" as a subjectivity produced through the lives, actions, and relationships of Indigenous communities. Indigenous practices, expertise, culture, and belief systems are pursued both inside and outside of capitalist production.[6] The location of Indigenous ontologies and practices both inside and outside of capitalist relations points to the importance of theorizing the conceptual categories outlined in this chapter at their spaces of overlap, and in the moments of fissure at categorical borders, rather than attempting a nomenclature that avoids these overlaps. These are rich spaces of contradiction.

"Indigenous," in the Canadian context, is a racialized category formed through the social relations of settler colonialism. Like other "minority" categories in Canada, Indigenous people are racialized in opposition to a white "norm." However, as Stuart Hall (1986) and David Theo Goldberg (1993) remind us, there is no general racialization or racism; rather, one must look to the histories and materialities that beget particular subject formations. In the context of Canada, and the area around Yellowknife, racialized minorities marked as "of elsewhere" are robbed of their indigeneity (Thobani 2007) – that is, their relationship to specific places – even while Dene, Métis, and Inuit people are defined by theirs. As Patrick Wolfe (2006) puts it, "so far as Indigenous people are concerned, where they are is who they are and not only by their own reckoning" (388). Here Wolfe is pointing to the territorial challenge Indigenous peoples pose to settler colonialism: identified through their relationship to place, by their very existence, Indigenous peoples are a danger to settler state sovereignty. The challenge is economic and political: Indigenous peoples' relational and communal approaches to land tenure undermine the bedrock of settler accumulation, private property, and settler state-building: sovereignty. Wolfe concludes that this challenge to the establishment and reproduction of the settler state/ economy has been the impetus for a logic of elimination inherent to the racialization of Indigenous peoples, a particular iteration of Fanon's

(1963) general observation that the racial formation of colonized subjects is forged in the violence of colonialism. I weave an analysis of the violence of the settler colonial racialization of Indigenous people into the chapters that follow, arguing that this racial logic, and its shifting iterations over time, have together shaped the uneven incorporation of Indigenous peoples into capitalist production in the mixed economy, as well as Canadian state activities targeting the social reproduction of Indigenous communities.

In particular, I am attentive to the gendered dynamics of settler colonialism that make Indigenous women's bodies the target for ongoing processes of settler violence (Smith 2005). In discussing contemporary manifestations of embodied violence against Indigenous women in Canada, Simpson (2016) writes that "Canada requires the death and so-called disappearance of Indigenous women to secure its sovereignty" (1). She explains:

> An Indian woman's body in settler regimes such as the US, and Canada is loaded with meaning – signifying other political orders, land itself, of the dangerous possibility of reproducing Indian life and most dangerously, other political orders. *Other* life forms, other sovereignties, other forms of political will. Indian women in the aforementioned example of the Haudenosaunee Confederacy transmit the clan, and with that: family, responsibility, relatedness to territory. Feminist scholars have argued that Native women's bodies were to the settler eye, like land, and as such in the settler mind, the Native woman is rendered "unrapeable" (or, highly "rapeable") because she was like land, matter to be extracted from, used, sullied, taken from, over and over again, something that is already violated and violatable in a great march to accumulate surplus, to so called "production." (7)

Indigenous bodies – especially Indigenous women's bodies – are a site of material and symbolic struggle, as evidenced by the ongoing crisis of missing and murdered Indigenous women in Canada.

Indeed, at an event discussing murdered and missing women in Yellowknife, Marie Speakman (2014), a victim services worker for the NWA/NWT, said, "I think about this as a woman and as an Aboriginal woman. I think about all the women vulnerable to this violence that has not been stopped. This is not just physical and sexual violence. Every day as Aboriginal women, we face systemic racism. We face prejudice and barriers in our work lives, our personal lives and on the streets."

Given the ways in which the violent racial formation of *Othered* corporealities has been central to the overt and "peaceful," or systemic, violence of settler colonialism in Canada, it is no surprise that some of the most detailed and compelling accounts of the embodied violence of colonialism to come out of academic institutions in Canada in the past twenty years have emerged out of the public health discipline.[7] As Fanon (1963) explains, colonialism is a process that seeps into the minds and bodies of both colonized and colonizer: the corporeal violence of colonialism is not confined to physical violence, but extends to the ways in which physiological and mental illness/health is produced and reproduced through pathologies of colonialism. In Canada, public health accounts critique colonial practices, such as the residential schools, the reserve system, and resource extraction, by linking these to varied and long-standing embodied repercussions. Kelm (1999), for example, writes of the ways in which colonialism has shaped Indigenous bodies: "Aboriginal ill-health was created not just by faceless pathogens but by the colonial policies and practices of the Canadian government" (xix). A health perspective is a useful reminder for the political economist to consider the different ways in which violence can target the body. In an analysis of extractive industries, this is particularly pertinent. Indeed, in interviews, discussions of the impacts of diamond mines often moved fluidly from concerns about domestic violence to problematic substance use to the disappearing caribou herds to conspicuously rising rates of cancer in communities. These concerns, while distinct, are experienced simultaneously, relationally, and corporeally, and will be approached accordingly.

At the same time, "Indigenous" is not just an identity ascribed by the Canadian state, or any state. While the broad grouping of Indigenous peoples has emerged as a result of settler colonialism in Canada, Indigenous identity has also been taken up to denote shared decolonizing practices, shared colonial pasts and presents, and shared values (Kino-nda-niimi Collective 2014). In this book, when speaking about specific histories, or particular groups or peoples, I use the terms, Dene, Métis, or Inuit, as these three groups are represented in the area around Yellowknife, and research participants self-identified as being a part of one or more of these groups. When I speak about *general* experiences in relation to the diamond mines shared in the region, I use the term Indigenous people, acknowledging the distinct histories and relations of Dene, Métis, and Inuit people.

It is also worth clarifying that writings on colonialism and autonomy in the North include two distinct, though interrelated, approaches to these terms: for some, the NWT and its populations (both non-Indigenous and Indigenous) are internal colonies of Canada, and the devolution of powers to the territorial government represents a move toward northern autonomy.[8] For most Indigenous communities, however, the government of the NWT (GNWT) represents another settler administration and the struggle is for *Indigenous* self-determination, regardless of what level of government this power is wrested from. I am speaking about colonialism – specifically, settler colonialism – in line with the second understanding. In this book, settler colonialism refers to ongoing processes of dispossession (Coulthard 2014) for the complementary – and at times indistinguishable – purposes of state-building and capital accumulation. I am concerned with the colonial continuities that tie the diamond-mining regime to processes of dispossession across space and time, and the ways in which that regime represents a state-sanctioned imposition of capitalist accumulation that impedes the social relations of Indigenous people. I do not use the term colonialism to deny the agency of Indigenous peoples.[9] Nor do I use it to obscure the ways in which Indigenous peoples and northern settlers have worked together to build local social, economic, and political structures (Southcott, ed., 2015), or the ways in which Indigenous peoples have participated in the political and economic architecture of the diamond-mining regime. Rather, I use the term settler colonialism to emphasize the struggle against the totalizing impulses of capital and the Canadian state in the regional mixed economy, and the structural and embodied violence of that struggle.[10]

I do not assume that settler colonialism advances in a linear way, with the North becoming increasingly "settled" over time. Rather, I aim to disrupt the teleological temporal lens through which Western political economic development views itself by approaching the relations of the northern mixed economy as informed by Tł̨chǫ cosmology. Tł̨chǫ cosmology, recall, is organized in terms of eras marked by ruptures and the forging of new relations between the Tł̨chǫ people and the land, animals, and other Indigenous and non-Indigenous peoples (Gibson MacDonald et al. 2014). Discussing her conversations with Tł̨chǫ elders, Allice Legat (2012) writes that the knowledge embedded in the stories of the Tł̨chǫ eras is transmitted with the intent of reproducing healthy, respectful relationships: "To know is to maintain proper respectful relations with all that is part of the *dé* ["land, ground, dirt, earth" – a living entity and constantly in flux (Legat 2)]" (18). This relational, place-based approach refuses Western temporal approaches to history and development that assume Western capitalist totality.

As a settler living outside the Tł̨chǫ Nation, and in keeping with Tł̨chǫ traditions around cultural knowledge holders, I do not discuss Tł̨chǫ cosmology in order to define or claim its teachings. Nor do I seek to generalize this knowledge to other northern Dene and Indigenous peoples. Rather, I engage with this rich way of knowing in order to unsettle linear, temporal approaches to development and honour the complex place-based relations of the northern mixed economy. The diamond-mining regime, like much extraction (especially in the neoliberal era), intensifies the inherent temporariness of capital accumulation. As I will make plain throughout this book, resistance to the temporary imperatives to orient production toward extracting value is found in the multi-scalar protection and reproduction of northern place-based relations. Centring the relational quality of community development helps reveal this resistance. Thus, in the chapters that follow, I analyse the shifting relations and the contested gradations of power relations between and within departments of production through an attention to the qualitative orientation of labours and their meanings. That is, how are labours performed and to what end? Throughout, I attend, in particular, to Indigenous women's complex negotiations of the competing imperatives that structure labour in the mixed economy, honouring the strength inherent to the day-to-day acts of resistance that reproduce northern place-based relations.

Conclusion

By analysing the social relations between and within departments of production at the level of social formation – that is, through the historical specificity of the contemporary mixed economy in the region in and around Yellowknife – the gendered impacts of the diamond mines at camp, in homes, on the land, and through community relations can be made discernable. I approach the mixed economy as a set of unfixed social relations wherein processes of colonization and decolonization are not expressed as either/or (either subsistence or capitalist production, either decolonization or colonization), but through qualitative and relational shifts. I employ this theoretical framework with the understanding that the analytical categories I use are necessarily incomplete and inadequate, but that by clearly articulating a defined analytical framework informed by specific political commitments and scholarly goals, my assumptions are laid bare and, thus, can be confronted, sharpened, and reshaped in the analysis that follows.

The theoretical framework outlined in this chapter identifies social reproduction as a site of de/colonizing contestation. The shifting ways in which social reproduction is performed in relation to both production for subsistence and surplus in the mixed economy is the primary site through which I explore the impact of diamond mining on northern gender relations. Indeed, the historical analysis in the following chapter demonstrates that, in the traditional land-based economy of the NWT, there is no distinction between social reproduction and subsistence in that both are forms of production directed toward the social needs of the community. We will see that social reproduction *becomes* distinct from subsistence through Canadian state activities that, on the one hand, racialize and marginalize subsistence, and on the other, feminize and restructure social reproduction toward a Western capitalist model. Today, social reproduction sits as a site of tension between subsistence and capitalist relations. It is through this expanded approach to production, then, that we are able to understand the material impetus behind the Canadian state's interventions into Indigenous women's activities related to social reproduction in the Yellowknife region, a theme that asserts itself again and again throughout the following chapters.

3 Wıìlıìdeh's Mixed Economy

I remember something I lived by, that is affiliated with the mines, when my uncle took me out when we were kids, out on the water to haul some water, we were out at old Fort Rae. Our auntie handed each of us kids a cup of water and he said, "You know, you won't always be able to drink from this lake. You won't always be able to drink this fresh water. You know, it tastes so good. You know why?" Because I was like, "Why? Why?" And he said, "Because some people are willing to give all this up for money. And that's why it's really important when you grow up to protect it. And that's what we're doing.'"

– Interview 116

Introduction

Having arrived at an expanded conception of production that encompasses social reproduction and subsistence, in this chapter I apply this framework to the political economic history of the Yellowknife region in the time that preceded the diamond mines. I challenge the frontier narrative that casts Yellowknife as simply a resource town by analysing the shifts in regional gendered and racialized relations within and between departments of production from the middle to the end of the twentieth century. In so doing, this chapter makes visible the violent shifts in gender divisions of labour arising from the development of the new, and newly racialized,[1] mixed economy as a result of northern settler colonial encroachments: specifically, the establishment of the gold mines around Yellowknife Bay and the Canadian state's activist role in restructuring Indigenous social reproduction in the NWT. Through these two colonial processes – that is, the encroachment on subsistence production through the imposition of extractive production, and the Canadian state's restructuring of social reproduction through targeted policies and punitive measures – a gendered division between subsistence production and social reproduction emerged in the newly mixed economy. By historicizing and denaturalizing the separation of social reproduction from subsistence, this chapter makes visible the central role of Indigenous women's labour in reproducing land-based relations in the region of study. This chapter, then, historicizes the theoretical contention advanced in the previous chapter that social reproduction is a locus of tension between capitalist production and subsistence.

My analysis unfolds in two sections. In the first, I examine gender relations in subsistence production in the traditional and early mixed economy of the Wıı̀lı̀ı̀deh region, suggesting that gendered divisions of labour in these economies are characterized by an interdependence in purpose and practice, with social reproduction and subsistence production "becoming" separate – and differently gendered – categories of labour largely through targeted policies pursued by the Canadian state. In the second, I take up the three departments of production that populate the expanded conception of production framework to analyse the development of Yellowknife's mixed economy throughout the twentieth century. I discuss the establishment of capitalist production in the form of gold mining and, later, government work, both of which were established through an ideology of "separate" settler/Indigenous development, though in practice they interacted relationally with existing subsistence production in the development of the regional mixed economy. From the mid-twentieth century onward, prior to the diamond mines, there were only limited attempts to integrate Dene communities into capitalist production; even so, the Canadian state took on an activist role in disciplining and restructuring Indigenous social reproduction to fit a Western, capitalist model. Canadian state activities disrupted the intergenerational social reproduction of Dene communities: a structural violence with past and present embodied consequences.

I. Gender Roles in Land-Based Economies

Subsistence, as I use the concept here, is labour rooted in a "grounded normativity" (Coulthard and Simpson 2016; Coulthard 2014), which in the NWT is performed for the purpose of acquiring or producing the needs of a household or community. While I am cognizant of the continuities between forms of non-capitalist, land-based labour around the globe (Mies and Bennholdt-Thomsen 1999, Kuokkanen 2011), I am referring to the *specific* ways in which subsistence is performed in the region of study, the sets of meanings the people performing this labour attach to it, and its racialization through settler colonial processes.

Throughout, I am attentive to the mutually constitutive relations between labour and land. Indeed, the YK Dene First Nation Advisory Council (1997) writes that

> since time immemorial, T'satsot'ine have stayed on the banks of the Wıìlıìdeh [Yellowknife River], coming to think of themselves as Wıìlıìdeh Yellowknives Dene and traveling throughout their traditional lands. As Indigenous people, Wıìlıìdeh Yellowknives Dene were born to their lands and in that sense are part of their lands. For Dene the "land" (ndeh in their languages) means about the same as the English word "environment": ndeh includes the soil and plants, the air and weather, birds, the waters and fish, trees, animals and people who use the land. From generation to generation, Dene are taught to respect the land because it is the source of their survival. Respect is paid in many ways: by using without damaging; by not wasting any part of animals, birds and fish; by offering to pay the land; and by learning to live with the land its changing without bringing change. (11)

This articulation of the land is a history, and a set of beliefs and material practices, a "grounded normativity" (Coulthard 2014: 60).[2] This groundedness links social relations as they shift through time.

In the traditional and, to some extent, the early mixed economy, subsistence and social reproduction were indistinguishable as they were performed with the same aim – that is, the provision of the social needs of the community. This interdependence is at odds with the Western capitalist feminization of social reproduction, with its nuclear, heterosexual household norm wherein social reproduction is structured outside of so-called relations of production (i.e., production for surplus) through the construction of a work/home divide. Thus, in Canada, there is a racialized tension between state-sanctioned nuclear social reproduction and the interdependence of labour woven through home, community, and the land in many Indigenous communities. This tension runs through the history of Canadian state/Indigenous relations – from the denial of women's leadership in subsistence practices to the racist surveillance of Indigenous motherhood (Baskin 2003) – and underpins the gendered labour restructuring driven by state intervention in the mixed economy in and around Yellowknife. I discuss the gendered relations of production of the Yellowknives and Tłıchǫ Dene, first in the context of the traditional northern economy in the region of study, and then in the context of the pre-diamond mixed economy.

George Blondin (1997) writes that the Dene were always in and of Dendendeh, a Dene term that means "land of the people," land that extends through and beyond the NWT. In traditional Dene economies, specific divisions of subsistence labour varied between different groups, but because of the geography and the climate of the NWT. However, they all shared a nomadic lifestyle wherein large game were most often hunted (usually a male task) and prepared (usually a female task) in cooperative parties (Asch 1977; Nahanni 1992; Irlbacher-Fox 2009). Phoebe Nahanni writes that "at the time of contact and generally to the 1920's the Dene traveled widely in family groups, periodically visiting the trading posts. Since then their movements gradually decreased in scope" (1992: 68). Like other Dene communities in the NWT, the Wıı̀lı̀ıdeh Yellowknives Dene lived nomadic lives, structuring their labour and their migration with the seasons. They would spend their summers gathering berries and medicines and fishing on the Yellowknife River, moving their camps north to hunt as the weather grew colder (Yellowknives Dene Elders Advisory Group 1997). The Tłı̨chǫ Dene camped up the north arm of Great Slave Lake in the summer (Zoe 2005).

Up until recently, the land use patterns of the Wıı̀lı̀ıdeh Yellowknives Dene brought them in and out of the land that is now called Yellowknife on an annual basis. Before the town and mine settlements, the moose and rabbit around the Yellowknife River provided sufficient resources for some families to stay year-round, if they chose; however, most spent their winters in the northern barrens. The Tłı̨chǫ Dene followed similar seasonal land use patterns. The YK Dene First Nation Advisory Council describes (1997) winters in the following way:

> Every member of Wıı̀lı̀ıdeh Yellowknives families who could walk in the barrens harvested wood, water, food, feathers, and wind-blown musk-ox hair. Women, children and old people who could no longer travel on winter trails collected berries, medicine plants, moss, lichen, seeds, fish eggs, and bird eggs. They set willow and babiche nets in lakes to catch fish and in shrubs to catch ptarmigan. They set snares and nets for winter fowl, and snares for rabbit and other small animals. Youth and adult hunters who did not have to stay with young children harvested large animals for meat and trapped larger fur-bearers for pelts and sometimes meat. (27–8)

While certain roles were designated for women and others for men, traditional Yellowknives Dene relations of production are characterized by interdependence between humans and between humans and non-humans and by respect for the land and all aspects of labour. Nahanni (1992), discussing gender relations among the Dene of the Dehcho region (west of Great Slave Lake), explains that "within the domestic unit, each member of the family performs tasks which were learned by observation or taught to members of the family. Division of labour for some tasks are gender related" (57). Like Harnum and colleagues (2014) and Usher and colleagues (2003), who approach subsistence production at an inter- and intra-household rather than individual level, Nahanni stresses the flexibility of these roles within household and inter-household groupings, with women and men taking on different tasks as needed (57–9). For example, after the men had successfully hunted caribou, Yellowknives Dene women used caribou hides to make tipis, blankets, floor mats, covers for sled frames, and other necessities for their households and communities.[3] They also made hunting cases and bags for the hunters (YK Dene First Nation Advisory Council 1997: 33). The YK Dene First Nation Advisory Council (1997) notes that the women took great pride in decorating the hunting cases and bags (34); their craftwork honoured both the hunters' work and their own skill.

Gina Starblanket writes that settler colonialism has required Indigenous communities to invoke traditional knowledge and practices "selectively and strategically in order to have the best chance of receiving protection against their infringement" (2017: 26). Western patriarchal power relations infiltrated Indigenous communities through, for example, missionary education that prescribed subordinated roles for women, trade relations that targeted men (Bourgeault 1983), and state political interventions that denied Indigenous women's sociopolitical status (Lawrence 2003), working in tandem with the dominant Western gender scripts through which traditional knowledge and practices were read contemporaneously. As a result, Indigenous women's subsistence roles have been marginalized. For example, the interdependence of gender roles in subsistence is obscured when Indigenous women's subsistence labour is described as "traditional culture" rather than as work (as noted by Irlbacher-Fox 2009), as with beadwork and sewing. In practice, beadwork and sewing are both traditional practices *and* work. Describing this labour as "culture" only marginalizes traditional labour as a "symbol" of anachronistic culture (McClintock 1995)[4] when in fact it is a form of production that contributes to the regional political economy. This same attitude naturalizes the feminized labour that is performed in sites of home, perceiving it as non-work. Indeed, settler populations tend to be fascinated by caribou hunting, which they view as requiring "grit," yet they pay far less attention to all the complex processes women undertake in preparing the same caribou for food, shelter, clothing, and household goods. Yet this "women's work" is vital for communities (Fanning 2018; Maracle 1988), both in terms of meeting their needs for daily and intergenerational social reproduction and in terms of the meanings and relationships associated with these activities.

When subsistence labour is implicitly conflated with typically male activities, such as hunting, it is through a gendered (by settlers) assessment of the *quality* of labour and the gender subjectivities of the people performing the labour. This characterization inaccurately imports a labour typology – informed by capitalist social relations – that feminizes and naturalizes much non-(capitalist) value-producing labour. The gendered nature of a particular form of labour and the subjectivities of the people performing that labour pose different questions than the labour's productive purpose. Settler ideological assumptions obscure Indigenous women's role in subsistence production: conflating subsistence labour with the types of work typically performed by Indigenous men draws on the settler capitalist ideological assumption that social reproduction is *separate* from capitalist production; in this mistaken parallel, subsistence production replaces capitalist production in the imagined social reproduction/capitalist production dichotomy. The division between capitalist production and social reproduction is a *real* division under capitalism insofar as capitalist labour is labour that, as we have seen, creates profit, whereas social reproduction is labour that provides the necessities for daily and intergenerational reproduction.[5] While, in practice, social reproduction and capitalist production operate in overlapping ways, they are defined by distinct purposes as well as distinct relationships to the dominant mode of production. In subsistence production, however, there is no such division: neither subsistence nor social reproduction is performed for the valuation of capital. Instead, both forms of production are oriented toward the social needs of the community. Capitalism has a contradictory dependence on social reproduction; by contrast, in the traditional economy, *inter*dependence of labours emerges from shared goals and shared benefits. Furthermore, no one form of labour is ideologically or materially subsumed by the other.

As we move to an analysis of the mixed economy, we must heed Starblanket's (2017) caution that characterizing gender relations in land-based economies as complementary can mask the Western patriarchal relations that have been imposed on Indigenous economies. In the next section, as we apply an expanded conception of production to the changing mixed economy of the twentieth century, we will see that a distinction between subsistence production and social reproduction materializes as interventions by private capital and the Canadian state racialize and marginalize subsistence production while attempting to feminize and reorient Indigenous social reproduction toward capitalist production.

II. Tracing Shifts in Departments of Production in the Mixed Economy

Recall that prior to the Second World War, while there were certainly settler/Indigenous socio-economic relations in the NWT, because of the dominance of land-based production, elders and scholars have tended to characterize the economy up until around 1945 as a traditional one. However, the gold mines established in Yellowknife in the 1930s began a turn toward prosperity in postwar development and would reshape the mid-twentieth-century regional political economy. The sixty years between Con Mine's first dig and the first flight to bring workers from Yellowknife to the diamond mines at the close of the century were ones of profound change for the Tłı̨chǫ and Wıı̀lı̀ıdeh Yellowknives Dene; however, these changes were not rooted solely or even primarily in the evolution of the extractive industry.

Scholars have characterized the imposition of capitalist production on Indigenous communities as uneven development; by uneven, however, they tend to be referring to Indigenous people's subordinated position within capitalist production.[6] I concur with this general characterization while also suggesting that, in the case of encroachments on northern Indigenous social relations, there was, above all, an unevenness *between* departments of production. During the gold-mining era, there was a semblance of separate economic development between settler and Indigenous economies. But this was a semblance, only, in that the gold mines were established through the dispossession of Indigenous lands and resources even while relying on Indigenous expertise of the terrain. That said, there were only limited attempts to engage Indigenous people in capitalist production as full-time wage workers in the gold mines.[7] Conversely, based on a strategic national interest in the North,[8] there was intense pressure for the *social* incorporation of Indigenous peoples into the nation-state. Thus, throughout the mid-twentieth century, the Canadian state targeted the social reproduction performed in Indigenous communities as it sought to remake Indigenous social relations in the image of white settler social reproduction. State initiatives, that is, approached social reproduction as feminized labour isolated from subsistence, and this racialized department of production sat uneasily with the increasing settler development of the regional mixed economy. An analysis of these incomplete and contested settler colonial processes makes it clear that Indigenous social reproduction became the locus of struggle between subsistence and capitalist production and that contemporary feminized responsibility for social reproduction in Indigenous communities (to the extent to which it has been internalized) has been the result of relatively recent interventions into northern Indigenous social relations.

a) Capitalist Production in the Mixed Economy

In the 1930s, the lure of gold brought a new set of social relations – a rupture – to the early mixed economy. While the fur trade is what initially drew Euro-Canadians to the NWT,[9] it was gold that enticed them to settle on the Dene land now known as Yellowknife. Mark Dickerson (1992) writes that "gold fever spilled over into the NWT from the Yukon [gold rush]" (18). In 1935, a gold deposit found on the eastern shore of Yellowknife Bay led to the development of Con Mine. Shacks were erected to house the mine workers in what is now Yellowknife's Old Town, a rocky shoreline that to this day is etched with frontier lore about adventure, living on the land, and escape from the ordinary (Hurcomb 2012). In historical accounts, the narratives vary, with the most significant variance found between those told by prospectors – involving the discovery of gold on an empty land – and the Yellowknives Dene account, which has it that two women picking berries for medicine found a rock full of gold and were convinced by a prospector having tea with their family to trade it with him for a metal pot (YK Dene First Nation Advisory Council 1997).

At first, the deposits at Con Mine, the first gold mine, were believed to be limited, and mine development was sidelined by the socio-economic imperatives of the Second World War. Young men who might have worked in the mines were being called on to fight; northern uranium extraction was prioritized over gold during the war (Dickerson 1992); and the North gained strategic importance for national security (both Canadian and American), which served to redirect state resources that might have gone toward northern extraction. War-related developments included the building of airfields and winter roads, as well as the construction of the Canol Pipeline to bring oil from Norman Wells, NWT, through Whitehorse to Alaska (Abele 2009a: 24), all on Indigenous land, thus disrupting caribou migration and flora and fauna (Barry 1992). However, after the war, prospectors found additional gold deposits at the Con Mine site and on the north side of Yellowknife, leading to the development of Giant Mine in 1948 and the renewal of the NWT gold industry. Over the decades, the two mines operated at varying rates of profit: both faced threats of closure due to increased costs of production and stagnant gold prices in the 1960s, but their prospects revived when the market value of gold rose in the 1970s. Throughout the mid-twentieth century, the mines provided relatively steady employment for many settler men in the new town of Yellow-knife. Con Mine would operate until 2003, Giant Mine until 2004.[10]

Initially, most workers were housed in camps adjacent to the mines. As the town of Yellowknife developed, mine employees and their partners and children moved to homes closer to the new town centre. Con Mine and Giant Mine were but a short drive from these homes. Histories of the gold mines describe hard labour and jovial settler and camp life; the proverbial stakes were set, and lives were built in the newest part of the "New World." For some workers, employment in the gold mines was the alternative to poverty in southern Canada, as is often the case with extractive labour in imagined frontier or "undesirable" spaces. As Jack Lambert, who worked in Con Mine in 1938, when it first opened, said, "When I was there, I don't think there was any turnover. People were coming in and if they got a job, they stayed there. This was the time of the 1930s, when having a job was something to be prized" (in Silke 2009: 14).[11] Meanwhile, settler and Indigenous miners (generally male, sometimes with female partners) arrived from other parts of the North – for example, seasoned northern prospectors and miners (Indigenous or not) arrived from Great Bear Lake, the home of the Dene community of Deline, some 500 kilometres northwest of Yellowknife and a site of silver and uranium mining. For its new residents, Yellowknife became a space of "temporary permanence," a local political economy offering stable jobs that could support families (Seccombe 1986).

Gold mining introduced a particular form of extractive capitalism to the Yellowknife area, one characterized by a Western male-breadwinner/female-caregiver model wherein the social expectation was that (presumably male) mine workers would be paid a "family wage" that could support a wife (responsible for household labours) and children. The gold mines were established without consulting the Yellowknives or Tłı̨chǫ Dene and operated without sustained efforts to employ Indigenous people; the steady jobs offered by the gold mines were filled almost exclusively by settler men. The gold mines – emblematic of models of extraction in the North for most of the twentieth century – took an approach to resource extraction that emphasized fixed capital and in so doing laid settler claim to lands that had already been marked by the nomadic patterns of Dene subsistence. As they developed mining towns like Yellowknife, the extractive regimes invested in the fixed infrastructure to support the mines themselves and the daily and intergenerational social reproduction of the workers. To make up these sunk costs, there was a material imperative to encourage steady employment and worker loyalty.

Ultimately, the mining town model only mitigated the fundamentally temporary nature of extraction. The violent collapse of the gold-mining regime in the 1990s (see chapter 5) and the rush to replace the industry with diamonds reveal that the social relations of the settler political economy built around the gold mines reflected only a "temporary permanence," one that fuelled the settler colonial drive to accumulate new physical and geographic spaces. At the same time, the presence of capitalist production within the regional mixed economy has been only sporadic, belying the Western settler colonial teleological ideologies that imagine the development of liberal capitalism as an *end result.* Instead, the ruptures in social relations caused by shifts in capitalist production have been mitigated by the longer-standing relations of subsistence.

For a time, though, the gold-mining regime offered material stability for many of the new settlers in Yellowknife, or for a certain subset of the new settlers: for male workers – generally *white* male workers – the gold mines offered stable, high-wage employment.[12] Those high wages were being offered partly as an incentive to attract workers to the North (Selleck and Thompson 1997: 6); however, the employment conditions were also shaped by the strong unions that took root at the two gold mines. Workers at Con Mine and Giant Mine first unionized in 1944 and became members of the International Union of Mine, Mill and Smelter Workers in 1947 (NWT Union of Northern Workers 2014).[13] As Selleck and Thompson (1997) note, Yellowknife was largely isolated from the southern union politics. Instead, the workers "learned to rely on themselves for contract negotiations, grievance hearings and health and safety meetings" (7). Particularly pre-1967 – when Yellowknife became the territorial capital and new home to a large number of government workers – union activists often led local politics and community activities, from holding mayoral office to running local radio programs and community groups (Selleck and Thompson 1997: 7).[14]

The gold-mining regime established a Western male-breadwinner/ female-caregiver model that largely excluded settler women from capitalist production while relying on their labour for daily and intergenerational social reproduction. The mines hired some settler women for administrative positions above-ground, but a woman did not work underground until 1981. As Silke (2009) notes, women's interest in working underground was met with pushback from male workers. For example, Pam MacQuarrie-Higden, who would be the first woman to go underground, began her employment with the mines in 1975 working in security. She described walking into the Con mineshaft headframe and watching the other miners as they "'just walked out' in protest of a woman entering the 'sacred' shaft building" (Silke 2009: A20).[15] While local settler women were kept above-ground – and Dene women and men were relied on for temporary labour, though largely excluded from stable wage employment – the mines recruited male settlers from the Canadian south. As the decades wore on, settlers racialized as "immigrant" began to fill employment positions at Giant Mine and Con Mine. For example, in 1951, as part of a federal program to recruit new immigrant labour to meet the needs of extractive operations, twenty-one Italian immigrants began working at Con Mine.

Government and private initiatives to hire "immigrant labour" speak to the ways in which extractive labour is structured through an employment regime that relies on transnational labour mobility but that also uses racialization and borders as means of hyper-exploiting particular worker subjectivities, a tendency that binds extractive processes across place and time (Galeano 1997). Thus, one must trouble the category of "immigrant labour" here, remembering that as an expression, it locates subjectivities inside or outside of the imagined national citizen (Thobani 2007) according to shifting social relations of a given time and place. While all of the settler workers were new to the North, and many had life or family histories that were fairly newly located in Canada, it was specific workers who, based on racialization, language, and social location, were deemed "immigrant." Based on interviews with people who immigrated to Canada in this time period to work in the gold mines, Silke (2009) notes that those workers who could not speak English were paid less than their English-speaking counterparts and that the mine culture was one that favoured assimilation to the dominant Anglo-settler culture. While there is an operational and safety explanation for why non-English speakers tended to work in lower-paying jobs, in extraction – as with other industries – language and education requirements tend to reproduce and reify racialized inequalities.

The gold mining industry was far more proactive and successful at attracting workers from southern Canada (whether the worker was perceived as immigrant or not) than it was at attracting northern Indigenous workers. Recruiting northern Indigenous people as full-time mine workers was not a predominant labour strategy for either of the gold mines, nor was it a goal of the Canadian state, apart from minor recruitment initiatives. The first survey on Indigenous employment in the northern extractive industry – conducted by what was then the Department of Indian Affairs and Northern Development – found that 3.4 per cent (14 people) of Giant Mine's labour force was Indigenous; 4 per cent (or 9 people) of Con Mine's labour force was Indigenous. By the 1980s, only 6 per cent of the NWT's extractive labour force was

Indigenous (Bone and Green 2003), yet more than half the territory's population was Indigenous (Saku 1999). And that 6 per cent included Indigenous workers from other parts of the North who relocated to Yellowknife to work in the mines; thus, it is an understatement to say that local Th cho and Yellowknives Dene participation in the gold mine labour force was minimal.

Yet it would be wrong to suggest that Indigenous women and men were not involved in the beginnings of extractive labour in Yellowknife. While it is true that few Indigenous women or men undertook steady employment with Con Mine or Giant Mine, the mine companies and early settlers relied on Indigenous people's knowledge of the local land and resources. From the very beginnings of settlement, Indigenous women were being hired to cook and clean for pay (Price 1974). Indigenous men were hired to do what was called bushwork – usually short-term operations that included preparing, building, and maintaining camp sites. Such work required knowledge of the land and expertise in skills like trailblazing, hunting, and building fires. Heather was a schoolteacher at the time, and her partner did occasional bushwork at the mines. She recounted that "jobs would come up and off they'd go. Sometimes, I remember, there were a group of them, like, all our friends, all went out for one winter ... And whenever they came back, they always had stories of adventure" (interview 105). In this way, the gold mining industry built a temporarily permanent regime with labour "from away," an ever-shifting settler subjectivity. This temporary mode of production was established in relation to, and in tension with, the place-based orientation of subsistence production in the early mixed economy (Abele 2015; Parlee 2015).

At the time, Canadian state efforts to engage Indigenous peoples in the NWT with the market economy largely concentrated on promoting the market exchange of traditionally subsistence-based activities (Rae 1976: 125–6).[16] However, from the mid-1960s onward, the gold-mining companies and the Canadian state began recognizing northern Indigenous peoples as a potential steady, local workforce – one that might be relied on to stay in the North, unlike some of the southern transplants. Government and extractive companies began exploring strategies for recruiting Indigenous people into mining; this included attempts to relocate Indigenous families from other parts of the NWT to Yellowknife to work at the mines (Collymore 1980). Unlike the relocation efforts tied to social reproduction (see below), attempts to incorporate local Indigenous communities into the mining workforce were largely unsuccessful. At the time, both the government and the mining companies characterized the problem as one of human resources, instead of coming to terms with the complete lack of Indigenous engagement or consent that characterized the gold mining regime. More to the point, the gold mines had been established on Dene land without any consultation, remuneration, or even acknowledgment.

Indeed, the Yellowknives Dene's account of the establishment of the mines and the town of Yellowknife contrasts starkly with settler narratives of exploration and discovery. The Yellowknives Dene First Nation Advisory Council (1997) write of the years preceding and during the first mine camps:

> The years following were sad ones for Dene. Surveyors and prospectors, eager to gain quicker access to the peoples' land, set fires that forced changes to migration patterns. Poisoned meat set out for fur-bearing animals resulted in untold deaths of sled dogs and people: an entire Wıı̀lıı̀deh Yellowknives community at Smoky Lake became victims of greedy trappers ... Many people died in a series of epidemics, the worst of which occurred in 1928, when an estimated 10 to 15 percent of the entire indigenous population of Denendeh died in six weeks during the summer. Wıı̀lıı̀deh Yellowknives survivors, fearing a return of disease in Wıı̀lıı̀deh-Cheh, stayed in the barrenlands year-round for four or five years. When they returned, they discovered newcomers in their traditional lands. (49)

Both the Tłı̨chǫ and the Yellowknives Dene had been engaged with settlers through the fur trade – indeed, Behchokǫ̀ was once Fort Rae, a camp established as a fur trading post. However, the founding of a settler town and the fixed capital of the gold mines had a markedly different impact from the fur trade. While the Tłı̨chǫ and the Yellowknives Dene were both impacted by the gold mines, the Yellowknives Dene felt the imposition more directly, as Yellowknife was established on the traditional land of their summer camps (see Image 2).

The Yellowknives Dene continued to engage in subsistence production, but with the new town, the new mines, and the new neighbours, their activities were deeply disrupted, both structurally and corporeally. Environmental contamination and settler land-use patterns drove animals from the area around the Yellowknife River, impeding the gathering, fishing, hunting, and trapping of small game. Caribou stopped coming to the area, forcing hunters to travel longer distances for large game. Because of the arsenic used to separate gold from the rock in mining operations, women were forced to change their gathering patterns: "Women used to pick berries in the area where uptown Yellowknife is now and in the Giant Mine area. The men used to portage to Long Lake

Image 2. Giant Mine on the shores of Great Slave Lake. In this postcard from 1948, the emissions from the mine site are visible (NWT Archives/Busse/N-1979-052:12947).

to hunt for caribou. We would set up camps to make dry meat and look for berries for the upcoming winter" (Sangris 1968/1972).[17]

   Like other Indigenous groups around the territory, the Yellowknives Dene were remarkably adept at adapting to the changes brought about by the settlers – they travelled farther for their winter hunts as the big game moved away from the mines and the town – but their resilience should not minimize the profound and violent rupture brought about by the attacks on the Dene land, resources, and bodies. For example, the arsenic that contaminated the water and the land led to the deaths of a number of animals and, tragically, the death of four children living in the family camps of N'dilo in 1951 (YK Dene First Nation Advisory Council 1997: 53).[18] The corporeal impacts of the gold mines' environmental contamination emphasize the tight links, and sometimes the indistinguishability, between structural and embodied violence as it relates to extraction and capitalist accumulation of new spaces.

Of course, the gold mining extractive regime would not be the only form of capitalist production introduced to the Wıı́lıı̀deh area in the twentieth century. Yellowknife is a town that defies easy categorization: a "resource town" to be sure, Yellowknife is also a government town. Thirty years after the gold mine opened, in 1967, Yellowknife was named the territorial capital, making government a significant source of employment there. Unlike the provinces, which brought a predetermined level of autonomy into Confederation, the NWT developed under federal administration through an evolving bureaucratic matrix in Ottawa.[19] Dickerson (1992) writes that from 1905 to 1921, the NWT was governed from Ottawa in an ad hoc fashion without a policy framework; then from 1921 to 1950, "a handful of civil servants in Ottawa ran the region as if it was their own bureaucratic fiefdom" (29).[20] However, the 1950s saw concerted efforts to develop regional self-administration in the NWT, including territorial representation in the federal parliament. This devolution led to the establishment of Yellowknife as the territorial capital and the corresponding move of administration from Ottawa to Yellowknife.[21] In 1953, the first mayor of Yellowknife was elected (Dickerson 1992: 18); in 1967, Yellowknife was named the territorial capital.

This regional self-administration was merely part of a long process of devolution that saw, arguably, one of its most significant developments in 2014 – the granting of territorial jurisdiction over subsurface rights.[22] However, in terms of the development of Yellowknife as a city and the introduction of new forms of employment, the move of the territorial administration from Ottawa was of deep significance. Hurcomb (2012) writes that "the Government of the NWT left Ottawa and arrived in Yellowknife, with an advance guard of 81 government employees. Almost overnight, Yellowknife changed from a mining town into a government center, with all that entailed, including the construction of office buildings and apartment complexes in New Town to accommodate the new arrivals" (10).

And indeed, the abrupt "arrival" was more than just metaphor. Archived CBC Radio North footage from Yellowknife in 1967 reads:

We start with the arrival at the territorial government on September 18, 1967. Everyone is anxiously awaiting the arrival of the new residents of the NWT. The stewardess has opened the door and I can see the figures of some of the first arrivals. I see our Commissioner, Stuart Hodgson, followed by John Parker and Mr. Gamble, the Treasurer of the NWT ... I think I see a pet skunk coming off. I seem to see a cage. The skunk is just behind Mr. Gilchrist, the assistant to the Commissioner. The skunk is very small and is completely deodorized. I would be interested in seeing its first encounter with a raven. (CBC in *Outcrop* 2000: 180)[23]

For those Yellowknife residents who took pride in their frontier settler subjectivity, it was the beginning of a new time, indeed. Reflecting the dynamism of the northern mixed economy, the "temporary permanence" of the gold mining economy gave way to the diverse forms of wage labour that would come to characterize the Yellowknife region: government jobs, an increasing number of arm's-length service and community development positions (Southcott 2015), and, as a result of Indigenous organizing for self-determination, a proliferation of Indigenous governance positions. Indeed, the newly localized administrative apparatuses would come to shape, and in turn be shaped by, the Indigenous governance regimes that would be built throughout the territory from the 1970s onward through Indigenous resistance, land claim negotiation, and struggles for self-government. This new, and newly differentiated, access to capitalist wage labour has shaped the gendered and racialized impact of diamond mining, discussed in the chapters that follow. However, while government and gold mining have jointly dominated capitalist production since the 1960s, throughout the mid-twentieth century it was Canadian state programs and policies that most profoundly impacted the subsistence and social reproduction of Indigenous people in the Yellowknife region.

b) Restructuring Subsistence and Social Reproduction in the New Mixed Economy

Prior to the mid-twentieth century, Indigenous peoples of the North had been somewhat exempt from the settler social interventions that Indigenous communities in southern Canada had experienced so intensely. As Philip Blake told the Mackenzie Valley Pipeline Inquiry, "For a while it seemed that we might escape the greed of the southern system. The north was seen as a frozen wasteland, not fit for the civilized ways of the white man. But that has been changing over the past few years" (Blake in Watkins ed. 1977: 6). The gold mines impacted Indigenous lives and labour in the area through processes of accumulation and, largely, exclusion; but it was the Canadian state that set out quite deliberately to restructure Indigenous social relations – specifically, by separating social reproduction from subsistence and restructuring the former to mirror that of Western nuclear families. The boundary that developed between social reproduction and subsistence production is, to be sure, socially constructed, mutable, and changing.

Subsistence production was arguably most affected by relocation initiatives, but it was simultaneously constrained by the state's "environmental conservation" initiatives. In its mid-twentieth century northern environmental programs, the federal government viewed its relationship to the land through a paternalistic conservationist lens, in contrast to the interdependent relationship the Dene had long nurtured with the land, plants, water, and animals. For example, in the 1940s, the federal government developed an initiative to poison wolves as a means of protecting the northern caribou. This project killed a large number of fur-bearing animals in the Yellowknife area. According to the YK Dene First Nation Elders Advisory Council, "Coyote suffered so greatly that they were soon extinct in Wıìlıìdeh-Cheh; the last one was seen in the area in 1979" (1997: 51). This brutally interventionist approach to "protection" is antithetical to the Dene's measured approach to hunting and trapping, which has evolved over centuries, is rooted in expertise and relationships to the land and animals, and is based on respect and a desire to sustain healthy relationships between people, land, and animals (Zoe 2005; Legat 2012).

Furthermore, as the city of Yellowknife grew throughout the 1950s and 1960s, and as its people set down roots, recreation activities in the area became more popular, leading to the disruption of sacred sites and traditional hunting camps. The parks that were established at the time did not protect the land for Indigenous use, but instead regulated and restricted subsistence activities within their boundaries (YK Dene First Nation Elders Advisory Council: 1997: 52). Those parks spatialized a settler approach to environmental conservation and wilderness recreation, and that same perspective would eventually shape the diamond mines' approach to subsistence. *Contra* nomadic patterns of land use, the state declared separate "conservation" areas and "residential" areas. In doing so, it mobilized its resources and coercive powers to relocate northern Indigenous populations to permanent settlements. Much as with the "temporary permanence" of Yellowknife's gold-mining settlement, the drive to create "permanent" communities is an example of the time/place tension at work in the mixed economy, and the contradictions therein.

Northern relocation initiatives were possible because of the collusive system that linked welfare payments to relocation/settlement and mandatory education. Government schools were made mandatory for Indigenous children; at the same time, welfare payments were made dependent on children's attendance. Indigenous people had to relocate to permanent settlements in order to access to those payments as well as social services. In the case of settlements adjacent to towns like N'dilo,

Dettah, and Behchokò, permanent relocation offered the significant incentive of allowing one to keep one's family together, for the children
could attend day school in Yellowknife. These processes positioned Indigenous social reproduction on the front line in the relationship between the Canadian state and Dene communities. The Dene had lived a
nomadic existence for centuries, moving from camp to camp based on
the seasons and the community's needs. The push for the Dene to settle
permanently in communities, rather than continue their nomadic subsistence production, living in different camps at different times of year
and moving flexibly based on the needs of the community, the climate,
and the status of the land and animals, was not a simple or immediate
effort (nor should it be construed as an entirely successful or complete
process, given that Indigenous peoples all over the North continue to
go out on the land in varying forms). Rather, the mass relocation came
about through a combination of factors: the impact of extractive projects and settlements on subsistence, various forms of direct coercion,
and strategic education and welfare policies.

Wilf Bean (1977), a former federal employee in the northern bureaucracy, discusses the professional incentives bureaucrats had for
"convincing" Indigenous people to abandon their traditional hunting
camps and nomadic lives, noting that his predecessor received a major promotion for "finally convincing the Perry Island people to move
to Cambridge Bay" (131).[24] This "convincing" involved directly tying
welfare payments to the new permanent communities. Given the racialized colonial tropes that critique Indigenous peoples for their reliance on welfare, it is important to clarify here that such payments
were not a welfare state social assistance program responding to poverty, but were instead treaty payments distributed in accordance with
negotiations between the federal state and northern Indigenous peoples (YK Dene First Nation Advisory Council 1997; Shewell 2004; Zoe
2005). Because of the diminished market value for furs at the time,
and the growing use of market goods in subsistence production, many
Dene relied on treaty payments (Coates 1985: 192–3). Thus, condi-
tioning these payments (reimagined as state "charity") on permanent
residence was profoundly coercive. As the Yellowknives Dene First
Nation Elders Advisory Council writes: "In 1959, when the Indian
Agent and RCMP made their annual visit to Dettah to give out Treaty
payments, they told Chief Joe Sangris to tell his people to come in from
their lands and stay in town permanently. The government wanted
the children to go to the church's school, he said, and the people had
to stay in a place where doctors and nurses could give them medical
attention" (1997: 53).

Some Yellowknives Dene families continued to travel to the barrens annually until the late 1970s, and Dene around the NWT continue to go on extended trips on the land; overall, though, the federal policy marked a major shift toward sedentary lifestyles for these communities. In the 1950s, there were around fifteen houses in Dettah and ten in N'dilo; both became permanent settlements for the Yellowknives Dene. Government-built homes in Dettah, N'dilo, and Behchokǫ̀ marked the beginning of a new reliance on subsidized housing with sliding-scale rents.[25] Unlike other small communities in the NWT, Behchokǫ̀, and especially Dettah and N'dilo, developed through their proximity to a primarily settler community, Yellowknife. As sedentary living impeded subsistence production by depleting the nearby lands of sufficient flora and fauna for hunting, trapping, and gathering, new forms of work opened up in government offices and the extractive industry. Thus, I argue that while Dene engagement in capitalist production increased at this time, that engagement was driven by the Canadian state's restructuring of social reproduction and subsistence, rather than the other way around.

As a result of government housing programs and the government subsidies for daily maintenance, Indigenous households found themselves in new proximity (literal and metaphorical) to the Canadian state, which demanded particular forms of household and kin relations. Dene households in the area, and Indigenous households more generally, found themselves being disciplined to conform to styles of household formation, care and education, common in Western states throughout the world and moulded around the heterosexual nuclear family norm, and thus conducive to the social reproduction of capitalist social relations. Social reproduction was being (and continues to be) restructured to address the needs of capitalist accumulation and Western social norms – *contra* subsistence land-based relations. In this way, the prevailing gender ideology attached to the capitalist economy – that is, masculinized capitalist production and feminized social reproduction – was imposed on Dene communities through Canadian state policies, albeit in contradictory and incomplete ways.

New state social services targeted the mother as the person solely responsible for the child. This meant that treaty payments were administered under two assumptions: that social reproduction reflected Western-style nuclear family arrangements, and that it was the woman's responsibility. Thobani (2007: 126) argues that mid-twentieth-century Canadian welfare state social policies were characterized by racist norms that targeted Indigenous and non-white women, along with their households and communities. These norms shaped welfare policy for the benefit of those mothers who fit white settler heterosexual and class-ed norms – either women read as "white" or, as Audra Simpson (2016) puts it, "aspiring towards an unmarked center of whiteness" – and discipline and surveillance toward mothers of colour, including Indigenous mothers. Thobani writes, "In the welfarist national imaginary, Native families were deficient; Native mothers deviant and a menace to their own children; and the nation the caring benefactor of these children" (109). In most northern Indigenous communities, while women are often the ones responsible for meeting the immediate needs of young children, children are seen as the responsibility of the entire community, and their care and teaching is approached collaboratively (Harnum et al. 2014). The Canadian state's erroneous assumptions about women's responsibility for social reproduction, centred in the household (rather than the community), have shaped its interventions into and surveillance of child care in Indigenous communities. Indeed, the state-driven restructuring of social reproduction laid the groundwork for the male-breadwinner/female-caregiver household model that would be imposed by the FIFO diamond-mining regime, discussed in the following chapters.

c) NWT Residential Schools

The racist devaluation of Indigenous parenting – specifically, Indigenous mothering – was institutionalized through residential schools. In the NWT, the push for permanent settlements was directly tied to residential schools and the so-called re-education of Indigenous children. In the immediate sense, some families moved to permanent settlements in order to remain together while their children attended mandatory government schools. In the longer term, children who attended residential schools were more likely to remain in towns, as they had been separated from subsistence education and relations. Residential schools operated in Canada with the explicit mandate of "civilizing" Indigenous peoples. In 1879, responding to a recommendation for residential schools, Prime Minister John A. Macdonald stated, "It has been impressed upon me as head of the Department that Indian children should be withdrawn as much as possible from the parental influence, and the only way to do that would be in central training industrial schools where they will acquire the habits and modes of thought of white men" (Miller 1996). "Civilizing," then, meant replacing community and family bonds with the attentions of strangers who abused the children, imposed racial hierarchies on them, and punished them when they tried to socially reproduce their own community – whether this meant speaking their own language, carrying on their religious practices, or telling the stories of home. Smith (2005), writing of the similar boarding school system in the United States, adds a gendered analysis to this "re-education." In discussing the Western forms of domestic labour taught to Indigenous girls, she argues that "the primary role of this education for Indian girls was to inculcate patriarchal norms into Native communities so that women would lose their place of leadership in Native communities" (37). These patriarchal gender norms stood in contrast to the interdependent gendered divisions of labour common to the children's home communities.

In the NWT, by the 1950s and 1960s, Indigenous children were being coerced into attending government schools, usually residential schools. In southern Canada by that time, residential schools had been operating for almost a century. Those in the NWT prior to the mid-twentieth century had been managed in an ad hoc fashion by missionaries. Between 1954 and 1964, however, the federal government opened five large public schools in the NWT, and flew children in from around the territory to attend them.[26] The goal was to place every school-age child in the NWT in a school by 1968 (GNWT Legacy and Hope Foundation 2013: 18). Coates (1985) writes that "by offering native children the standard Canadian curriculum, the government hoped to speed their assimilation into the broader Canadian society" (194). Yellowknife was home to Akaitcho Hall (now Sir John Franklin High School). The last residential school to close in Canada, Akaitcho Hall operated as a boarding school until 1996. It housed children from small communities throughout the territory; meanwhile, children from Dettah and N'dilo were able to attend day school as long as their parents remained in the newly "permanent" settlements. According to the GNWT Legacy and Hope Foundation,

> while the system was late in coming to the North, its impact was significant and continues to the present. A far higher percentage of the Aboriginal population in northern Canada attended residential schools than was the case in the rest of Canada. According to the 2001 Statistics Canada Aboriginal Peoples Survey, over 50% of Aboriginal peoples 45 years of age and older in the Yukon and the Northwest Territories attended a residential school. (2013: 20)

Both structurally and corporeally, residential schools were the most violent arm of the racist Canadian state project aimed at restructuring Indigenous social reproduction. The education and socialization of children – both aspects of social reproduction – were being restructured to accommodate Western capitalist social relations, yet there was no parallel effort to integrate Dene people into the modes of capitalist production emerging in the region at the time – namely, gold mining (recalling, here, that while local Dene and Indigenous people from other parts of the North did work for the gold mines, it was generally through short-term contracts, and there were no substantive efforts to hire Indigenous people for full-time mine work). Indeed, some felt they were being trained for jobs that did not exist, while being separated from the intergenerational education they would have received in their home communities through engagement in subsistence production (Kakfwi 1977). Traditionally, the entire community took part in educating children by including them in all parts of camp life: "Teaching and learning happened during the course of day-to-day family life. A child was never asked, 'What are you going to be when you grow up?' The answer was obvious and the training program was well-established" (GNWT 2013b: 29).

Efforts to recruit Indigenous employees into both government and extractive wage work would increase in the region in the mid-to late twentieth century, driven by state development policy rather than trends in extractive employment (Rae 1976). At the same time, residential schools shifted in substance over the decades. Among more recent students, the stories of abuse and overt colonial racism are fewer; but they too talk out dislocations, generational knowledge and relationship gaps, and feelings of "not belonging anywhere," which speaks to this day of the painful contradictions that Indigenous people must navigate in achieving educative or Canadian state-sanctioned "success." Kakfwi (1977) notes, from his own experience, that Indigenous children were being taught how to be capitalist subjects. The loss of linguistic, cultural, and relational learning when education is separated from subsistence production and community relations has been well-documented, most recently by local participants in the Truth and Reconciliation Commission (2015), and was expressed by a number of research participants in interviews and talking circles.

The residential school program failed to substantially integrate Dene people into the wage economy. It did, however, succeed in rupturing local social relations by separating children from their parents and communities, their language, and the intergenerational learning necessary for the reproduction of land-based economies. Indeed, impeding the social reproduction of Indigenous communities had been the original intent of residential schools, as expressed by public works minister Hector Langevin when he announced the residential school plan to the House of Commons in 1883: "In order to educate the children properly we must separate them from their families. Some people may say that this is hard, but if we want to civilize them we must do that" (in GNWT 2013b: 15). Though the language justifying residential schools may have changed over the years, the original intent has stayed the same, and its impact has had devastating and violent intergenerational impacts, and still does.

The waves of residential school trauma mingle with contemporary assaults on Indigenous social reproduction, including the new, and newly gendered, strains imposed by the diamond mines (see Part III of this book). In an interview, Martha, a community worker, noted the temporal proximity of residential schools in the NWT:

> We're talking about people who are alive today who were in residential schools, some who were in the first residential schools. So this is all recent. Very. And there's a really big disconnect and I don't think any [diamond mining] corporations take that stuff into account. Because they're like, "oh no, it's [mining] been done forever." And it hasn't. In a span of someone's life, it's only a few people. A few people having kids that we're talking about. And I think failing to take those things into account, it's created a lot of jobs, a lot of education, but socially there's going to be a lot of ripple effects to it. (Interview 202)

Martha's point reminds us how quickly shifts in social reproduction in the region have occurred over the past century. Today, most of the people who attended residential school are middle-aged or older, though there is no easy demarcation of an "end" to the schools (Akaitcho Hall did not officially close until 1996). Indigenous people who attended residential school, or whose parents or grandparents did, often express the experience as a foundational and devastating factor in their lives and relationships. Though the interview schedule I used did not include questions about residential school, research participants frequently raised the subject. Christensen's (2017) field research on Indigenous homelessness in Yellowknife and Inuvik delivered a similar finding: Christensen named residential schools as the colonizing process with the strongest impact on the people with whom she spoke. Research participants in this project explained that residential schools broke interpersonal, community, and cultural ties. Children were separated from their parents, elders, and communities and taught that their own people's ways of being, knowing, and learning were wrong. When these children became parents themselves, many felt they lacked the intergenerational tools and skills required to build a home. Della Green, whose mother attended residential school,[27] expressed the complex and sometimes contradictory experience of being raised by a residential school survivor, and the difficult labour her mother performed as she engaged in the daily and intergenerational reproduction of a home and community she had been taught was wrong. She shared this story:

> I was preparing a presentation for the Residential School survivors, and my co-worker and I were searching for some photos to use for our presentation, and I came across this photo of six students. My heart stopped when I saw the photo, as I recall my Mother having a photo on her wall in the dining room of her home. It was of the six oldest of her children (that included me). It was actually almost the same as the one I saw on the computer screen.
>
> All three girls [in the residential school photo] had the same haircut as the photo [in my dining room], and we all dressed the same. The boys all had checkered shirts on, and blue jeans with suspenders, and they also had the same haircuts in the photo. I asked my husband to send me that photo, as I recall him taking a photo of that photo at our Mother's house, so he did. And sure enough, there we were! We could have been the same students in that photo!
>
> I knew right then and there, even though my Mother said she was not affected by the Residential School, she dressed us just like the students in that photo! I felt sad and emotional about the whole ordeal, for my Mother who is gone now. I often wonder if she really knew. (Green 2015)[28]

Della's story illustrates the structural violence that residential schools inflicted on day-to-day and intergenerational Indigenous social reproduction. Yet at the same time, she characterizes her mother, not as a passive recipient of this violence, but as an actor navigating her roles and responsibilities and drawing upon the resources she had available to her as she faced the de/colonizing processes of social reproduction.

Maria Yellow Horse Brave Heart (2003), Hawkeye Robinson (2006), and Christensen (2014) link analyses of the impacts of residential school trauma to long-standing experiences of colonial trauma: a "collective emotional and psychological wounding that occurs over the lifespan and across generations" (Christensen 2014: 811). And indeed, participants in this research often expressed fluid experiences of residential school and other forms of settler colonial violence. This fluidity extended to relational interpretations of the impact of residential schools and the impact of the diamond mines. For example, when discussing the diamond mines, Jenna, a Dene woman living in one of the communities just outside Yellowknife, questioned whether she was describing the effects of residential schools or the diamond mines, moving back and forth on the matter in conversation:

> They always blame residential school. You know, [for] drinking and what not. And I'm not putting it down. But maybe they don't see the whole colour or the whole picture. But some of them don't realize it's mining, too, that affects them. Because, you know, they're working, they're working and then they save enough money to put a down-payment on a house. And then they get laid off. And they were getting a chunk of money. Like, sometimes, I'd be getting $1800, roughly $1800 every two weeks when I was doing four and three. So can you imagine what they're getting? The ones who work underground? A lot more than me. Maybe $3600. Every two weeks. Yeah. And then whoosh, you get laid off. And maybe it's affecting them, but they're not really sure. And here they say residential, residential, but who knows? Maybe it's the mine. You just can't always blame residential. I mean, it did impact a lot of people. I don't put it past them. It might be just that, too. But, you know, it really affected a lot of people. It's so common. It's pretty sad. (Interview 109)

In articulating a blurred boundary between the impact of residential schools and that of the diamond mines, Jenna is expressing a revealing ambivalence regarding Dene relations of production: time is collapsed as different experiences of settler colonialism act upon the same place and its peoples. In some ways, Jenna is echoing Martha's contribution, in that she does not see the impacts of the diamond mines and the impacts of residential school as easily separable.

However recent, residential schools are no longer the order of the day. Today, Indigenous children attend public day schools run by the NWT government. While it is beyond the scope of this book, northern public schools remain a rich site for analysing community-level social reproduction. As Heather MacGregor (2010) notes, since the 1970s, important decolonizing gains have emerged from decades of work by Indigenous peoples and their allies working both outside and within northern government offices and schools. In the NWT, the public school curriculum now includes significant Indigenous content, with community elders as teachers (see, for example, GNWT 1993, 1996, 2012, 2014c). At the same time, in northern schools, tension remains between the public school model – focused as it is on preparing students for engagement in Western, capitalist society – and the transmission of local Indigenous knowledge (MacGregor 2010: 24).[29]

The apprehension of Indigenous children by child services has arguably taken the place of the "new residential school" (Blackstock 2016), a term thrown around with the casual gravity that comes with a story told too often, across too many generations. The "new residential school" refers to the staggering rate at which Indigenous children are taken from their homes by the Canadian state and placed in foster care. Christensen (2014) writes that in Canada today there are more Indigenous children in foster care than there ever were in residential school (812). Nationally, First Nations, Inuit, and Métis Nation children make up 7.7 per cent of the population under fifteen but a devastating 52.2 per cent of children in foster care in private homes (Government of Canada 2018a). For example, in the NWT in the 2012–13 fiscal year, 1,042 children received services from Child and Family Services. Of these, 372 received voluntary services aimed at mitigating child protective concerns, and the other 670 received other forms of "protective services" (i.e., they either were removed from their caregivers and placed under the temporary care of the Director of Child and Family Services [266] or were taken under some other form of protective services) (Office of the Auditor General 2014). Given that 95 per cent of children receiving services are Indigenous (Office of the Auditor General 2014), the 670 children under protective services and the 1,042 receiving services in general represent a full 3 per cent and 5 per cent respectively, of the approximately 22,000 Indigenous people (GNWT 2015a) living in the territory.[30]

There is clear continuity, then, between the systemic racist interventions into Indigenous social reproduction in the mid-twentieth century and the levels of apprehension today. The surveillance of Indigenous households and the apprehensions of Indigenous children are wrapped up in both long-standing settler colonial ideologies and contemporary material inequality, which, as we have seen, is also colonial. Part III of this book examines the relationship between the diamond mine and inequality in the northern mixed economy. Madeline, a white northern woman who worked in both government front-line and management positions, described the discrepancies in how social services approach non-Indigenous and Indigenous households. Noting that most government social service workers are non-Indigenous and come from southern Canada, she described the challenges and limitations workers face in engaging with, and assessing, care in Indigenous households:

> So right now we don't have anything forced into our system about how to deal with First Nations families. So it really relies on each worker's

approach to decide to build those networks ... People talk about community standards all the time, but you walk into the house and ... you're trying to be culturally appropriate and you don't even know what you're supposed to be looking for. And when is it poverty and when is it neglect? (Interview 302)

Madeline here is reflecting on the difficulties in assessing observations of poverty and concerns about neglect or safety in the context of a settler colonial program working with Indigenous populations with unique ideologies and structures of child care and child-rearing. Certainly, it is difficult for front-line workers in child protection services to provide thoughtful and appropriate services that prioritize children's safety while at the same time valuing the integrity of care relationships, and in both non-Indigenous and Indigenous homes, child protection can manifest itself as the punishing of poverty. And certainly, given the significant levels of household income inequality in the NWT, this is part of the northern child protection dynamic. That said, there is a racialized tension informing this concern, given the long-standing history of Canadian state interference in Indigenous households, the racialization of poverty in the region of study, and the disparity between Western and Indigenous norms of "home."

Conclusion

The NWT's mixed economy is characterized by a gendered settler colonial hierarchy that has made Indigenous women's lives, bodies, and labour a site of surveillance and tension. The reorientation and restructuring of Indigenous women's labour under the diamond-mining regime are occurring in the context of a social landscape that has undergone rapid-fire shifts in social reproduction and production since the mid-twentieth century. The mixed economy is also characterized by a material resilience and cultural wealth that has emerged from its place-based subsistence orientation, as well as by the creative, unique ways in which that economy has developed over the past century (Southcott, ed., 2015). In this landscape, social reproduction has emerged as a dynamic site of de/colonizing tension in the shifting mixed economy.

In discussing the diamond extractive regime in the following chapters, and particularly in Part III, it is worth noting the relative novelty and tenuousness of settler social relations in Yellowknife. In many ways, Yellowknife is a town that was born nostalgic, residents warily eyeing newcomers a few years after they ceased to be "new" themselves, and each generation crying out for the authenticity of the decade past. Within a very short time, Yellowknife moved from mine camp to mining town to government town, in this way establishing a "temporary permanence." In defiance of history, there was a local social assumption that things would always remain as they were. This assumption proved to faulty: as the gold mines ceased producing adequate profit margins in the 1990s, the mines turned to intensified labour exploitation, resulting in strikes, community unrest, and violence. It is in this context – the context of socio-economic rupture – that the diamond mines emerged.

PART TWO

The Political Economy of Diamonds

4 The Global Political Economy of Canadian Diamonds

Icy and pristine, our Canadian-mined diamonds are called Arctic Brilliance. A stunning celebration of your heritage, these styles are distinctly, decidedly Canadian.

– Peoples Jeweller (2019)

The development of the Canadian diamond industry was, in many ways, in pursuance of the exceptional, or more to the point, an effort to *be* exceptional. I am referring not to the quality of the gemstones, though it is often described in similar terms. Rather, I am referring to the global political, economic, and cultural processes that have differentiated Canadian diamonds from unsavoury "conflict" diamonds. Schlosser (2013) writes of the "nation-branding" that has shielded Canadian diamonds from the quiet enormity that is the global Canadian mining industry and the powerful narratives that invisibilize the labour processes and social relations of extraction, as though diamonds emerged of their own volition from the Arctic ice, as this chapter's epigraph would suggest. Disguised commodity chains are a fundamental part of luxury goods production (Hartwick 2000); yet Canadian diamonds are marked as exceptional *because of* their relationship to the site of extraction. Imagined "Canadian-ness" imbues NWT diamonds with an assumed environmental and social responsibility. This national essence is infused through natural imagery (maple leaves, polar bears) literally branded upon the diamonds, and, for the more discerning ethical consumer, stories of modern employment brought to remote northern Indigenous communities.

However, as one research participant noted, "[diamonds] are said to be a girl's best friend. I'm not sure which girls they are because it's certainly not anyone in here" (Talking Circle 2014). As we will see in the following chapters, the majority of diamond revenues flow south, and the images of natural northern beauty and happy northern workers obscure the socio-ecological impacts of the diamond mines that thread through departments of production in the NWT's mixed economy. In this chapter, the first of two providing the political-economic context for the empirical analysis of Part III, I widen the lens of analysis, positioning the Canadian diamond-mining industry in the global political economy of diamonds. I begin with a brief history of diamond mining and then shift focus to the controversies over conflict diamonds in Sub-Saharan Africa in the 1990s and Canada's role in the international regulatory response to those diamonds. I discuss how the (presumed) ethical Canadian diamond mines were established as a way for Western interests in the diamond industry to distance themselves from the stain of "blood diamonds." Indeed, while the devastation wrought by African conflicts funded and at times fuelled by diamonds cannot be overstated, in this chapter, I trouble the category of conflict mining, which characterizes violence as an aberration from "normal" extraction. That aberration is coded through racial capital, wherein local African politicians and miners are cast as deviant, while global Western players like De Beers are able to hide behind the presumed respectability of law-abiding Western capitalism. The snowy white landscapes of the Canadian diamond mines – a racial metaphor too on-the-nose, yet widely popular in the industry – has come to represent the path to responsible consumption, the way forward for the diamond-as-love story that made diamonds the popular luxury commodity they are today.

I. The History of Diamond Mining

The dominant narrative of diamonds is infused with imaginations of riches, singularity, and even mysticism. Diamonds are the purest form of carbon, forged over billions of years. The first diamonds were mined in India, and they have adorned elite figures throughout Western history, from Marc Antony's robe at Cleopatra's coronation to the crowns of medieval European royalty (Smillie 2010, 29). The early diamonds were rough stones that bore little resemblance to today's diamonds. Cutting and polishing techniques were developed during the early Renaissance, revealing gems with a unique ability to refract light. It was the human labour of cutting and polishing that created the luminous stones we know today, just as it was it deliberate human intervention that granted diamonds the designation of "rare" luxury commodities.

While diamond mining stretches back to antiquity – the earliest diamond mines operated in India as far back as 800 BCE (McCarthy 1942)[1] – it was the development of the De Beers diamond operations in South Africa in the late 1800s that established the modern diamond industry and its accompanying commodity characteristics. Ian Smillie, retelling the legend of diamond discovery in South Africa, writes that in 1867, a fifteen-year-old South African boy named Erasmus Jacobs happened upon a 21 carat diamond – aptly titled the "Eureka Diamond" – inadvertently spurring the South African diamond rush (Smillie 2010: 30). Following the discovery, a number of diamond operations were quickly established in the region. One of the most profitable deposits was under a farm owned by two brothers named De Beer. In 1880, a young Cecil Rhodes, already invested in diamond mines, purchased the operations on the brothers' farm and formed what in 1888 would be named De Beers Consolidated Mines. By 1890, De Beers controlled over 95 per cent of global diamond operations (Smillie 2010: 36–7).

This monopoly approach – specifically, the De Beers monopoly – would characterize diamond mining, production, and distribution for almost a century. Rhodes used his control over global diamond production to promote the perception that diamonds were a rare luxury commodity: he tempered production and controlled distribution. Smillie writes that "[Rhodes] decided that there should be a single channel for the distribution of diamonds, and he worked with London merchants to create a syndicate that would take everything his mines produced, controlling the price as the diamonds moved further along the pipeline toward the consumer" (2010: 36–7). Smillie contends that smaller diamond operators benefited from De Beers's control over the market because it kept diamond prices high. For that reason, De Beers faced little corporate competition or pushback against its monopoly. De Beers maintained this control for much of the twentieth century, and while the market has since diversified, it remains the leading corporate entity in the diamond industry.

The diamond industry would not be what it is today were it not for the famous De Beers marketing scheme, "A Diamond Is Forever," the slogan written in 1947 by Frances Gerety, a New York copywriter. In 1999, *Advertising Age* declared it the best slogan of the century (in Bell 2013: 190). More than a catchy jingle, "A Diamond Is Forever" articulated a shift in De Beers's marketing that would sustain the diamond industry for the rest of the twentieth century; indeed, it would change the rituals of romantic love in the Western world and, increasingly, around the globe. I am, of course, referring to the diamond engagement ring, an uncommon practice until the 1940s, which has since asserted itself as a mainstay of heteronormative Western marriage rituals. The De Beers campaign steered high-quality diamonds away from the luxury market toward middle-class North Americans and Europeans and, indeed, much of the world. For example, in Japan, the percentage of engaged women given diamond rings rose from less than 5 per cent in 1967 to 60 per cent in 1981 (Epstein 1982). Today, China and Japan follow the United States as the largest national diamond markets (De Beers Group 2018). For many, diamonds are less of a choice than an obligation, a heterosexual marriage ritual no less mandatory than a white dress or a wedding cake.

Diamonds and the narratives that surround them epitomize commodity fetishism as it plays out in the production and consumption of luxury goods. Commodity fetishism is Marx's term for the processes whereby commodities are separated from the labour relations that produce them and are themselves seen to possess the agency and energies inherent in these mystified labour relations – so, in our example, the consumption of the diamond is propelled by the gem's quality as a shining and eternal symbol of love rather by the physical and rhetorical productive processes that have made it so (Fine and Filho 2010). Today, however, the concept of commodity fetishism is often used as a means to investigate and explain the large-scale consumption of seemingly unnecessary and usually expensive objects, as well as the ways in which certain commodities are imbued with the quality of being "ethical" (Carrier 2010). Writing about luxury goods, Elaine Hartwick (2000) contends that "commodity signs, such as advertisements, must disguise the labor processes, chains of connection, and material conditions involved in producing and distributing commodities" (1190). Why do people line up outside Apple stores to purchase the new iPhone at exorbitant prices when their current cellphone still performs all of its requisite functions? Why does a shirt emblazoned with a small silhouette of a man playing polo sell for exponentially higher prices than a shirt with a similar cut and fabric lacking that same logo? And, for our purposes, why does the opening of a small box to reveal concentrated carbon elicit tears in eyes and lumps in throats, whether first-hand or voyeuristically through Instagram shots or the plenitude of films that use the diamond reveal as the apex in their story's arc?

I take Arjun Appadurai's (1986) point that luxury commodities are not a category opposite to, or wholly distinct from, commodities necessary for day-to-day social reproduction. Given that "consumption is eminently social, relational, and active rather than private, atomic, or passive" (31), the division between luxury and necessary commodities is social, context-specific, and ever-changing, as evidenced in the mid-twentieth century shift in diamond consumption from the elite to the middle class. In Marx's original formulation, commodity fetishism characterizes consumption, in general, under capitalism; in this regard, luxury commodities like diamonds intensify the significance of commodity as symbol. Appadurai writes that what distinguishes goods as "luxury" is that their "principal use is rhetorical and social, goods that are simply incarnated signs" (38).[2] Thus, even more than with commodities deemed necessary for social reproduction or as a means of production, there is a cultural imperative to distance luxury commodities from the labour relations whereby they appear in display cases and, eventually, palm-sized boxes. For an example grounded in different forms of extraction in Canada, consider debates around the Alberta oil industry. Methods of extracting bitumen from the Alberta oil sands are manifestly destructive. Aerial shots of the Alberta oil sands capture kilometres of decimated land, entire ecosystems laid waste. However, criticisms of the oil industry are countered by the central role oil plays in daily life, recalling, here, that the division between "necessity" and "luxury" is socially made and that the "need" for oil built into the structures of daily life is fought for and protected by the oil industry.[3] People do not buy oil as a representation of their aspirations; as such, it is relatively insulated from the pressures of public perception.[4] Diamonds, by contrast, are an industry built on little else.

Thus, as successful as the conflation of diamonds with the promise of everlasting love has been, diamonds, like any luxury commodity built upon its fictive aspects, are intensely vulnerable to shifts in public perception and culture. Unlike other valuable minerals, such as gold and silver, diamonds do not hold their value and are difficult to resell (Epstein 1982). It is no wonder, then, that De Beers made it through the twentieth century spending an average of $200 million a year on advertising (Smillie 2010: 41). Diamonds are vessels of desire, refracting not just light but aspirations of (hetero-monogomous) love, (capitalist) prosperity, and (Western nuclear) family. Encapsulating this slippery web of aspirations is no small feat, a project built upon the ideological masking of chains of production. It was at the end of the twentieth century – the century that linked diamonds "forever" to romantic rituals spanning countries, cultures, and class – that the diamond industry met its greatest challenge: revelations of horrific violence funded by the diamond trade in Sub-Saharan African countries, which threatened to permanently stain the "love" industry.

II. Conflict Diamonds

I contend that the Canadian diamond industry – in particular, its focus on "ethical diamonds" – was established in part through the country's privileged location to the global public and political reaction to conflict diamonds. However strategic the Canadian industry's relationship to conflict diamonds may be, this in no way diminishes the devastating magnitude of what came in the early 1990s to be known as "blood diamonds." The term "blood diamonds" refers to gems extracted by small-scale mining: those stones became the subject of conflict as well as a source of funds to continue it. They led to widespread violence and displacement in Sub-Saharan Africa throughout the mid-twentieth century. Bieri writes that "blood diamonds have fueled and funded wars, massive death and refugee crises in Angola, Sierra Leone, Liberia, the Democratic Republic of Congo (DRC) and Côte d'Ivoire. It is estimated that four million people have died in wars involving conflict diamonds over the past few decades" (2010: 1). Before discussing the specifics of these conflicts, I contextualize "blood diamonds" in the violent continuities that tie extractive projects together across time and space, challenging the framing of "blood diamonds" as an aberration from "normal" processes of extraction. The violence of the diamond industry must in fact be seen as part of the longer history of extraction, colonialism, and violence.

Perhaps more than anyone, in his expansive and meticulous prose, Eduardo Galeano has revealed the violent continuities that link global processes of extraction. In *Open Veins of Latin America: Five Centuries of the Pillage of a Continent*, he traces the bloody processes of colonial extraction of Latin American resources that funded early capitalist development in Europe:

> Latin American silver and gold – as Engels put it – penetrated like a corrosive acid through all the pores of Europe's moribund feudal society, and, for the benefit of nascent mercantilist capitalism, the mining entrepreneurs turned Indians and black slaves into a teeming "external proletariat" of the European economy ... The price of the tide of avarice, terror, and ferocity bearing down on these regions was Indian genocide: the best recent investigations credit pre-Columbian Mexico with a population between 30 and 37.5 million, and the Andean region is estimated to have possessed a similar number; Central America had between 10 and 13 million. The Indians of the Americas totaled no less than 70 million when the foreign conquerors appeared on the horizon; a century and a half later they had been reduced to 3.5 million. (1997: 38)

Galeano captures the devastating scope of early extractive violence in Latin America. While lands were torn up across the continent and local socio-economies were forcibly reoriented toward Spanish- and

Portuguese-led mining ventures, bodies were destroyed either through the overt genocidal annexation of lands and peoples or through the slow (and not so slow) degradation wrought by dangerous, punishing, and poisoning extractive labour. Extractive violence characterized European colonization throughout the mercantile period and early capitalism (Moore 2003) and has shaped the uneven development of the global political economy during the modern phases of capitalist development.

This violence has hardly gone unnoticed, and a number of theories have emerged seeking to explain the relationship between violence and resource extraction. The "resource curse," for example, places political economic agency in the resource itself, explaining poverty in a country through its mineral-rich soil or regional violence through the presence of oil reserves (as critiqued by Bellamy Foster and Clark 2004). By contrast, I locate agency within the social relations of extraction and theorize the violence of resource extraction through capital's relationship to land and resources, the "metabolic rift," and through the violence of colonial ideologies and political economies that organized and justified transnational extractive labours and commodity chains. The "metabolic rift" refers to the "'irreparable rift' in the 'metabolic interaction' between humans and the earth" (Marx in Bellamy Foster and Clark 2004), wherein the earth is objectified as extractable object and exploited to the fullest extent possible for the pursuit of capital accumulation. This destructive distantiation is in contrast to – and also an ongoing threat to – the interdependent relationship to the land practised by the Dene in this book's region of study, as well as by Indigenous people throughout the world: recall what Coulthard (2014) calls a "grounded normativity." While all capitalist relations can be characterized by the violence of the "metabolic rift," resource extraction is a form of production that is spatially and temporally specific: it is spatially specific in that it requires control over and often destruction of spaces to fulfil its purpose; and it is temporally specific in that it intensifies the inherent temporariness of capitalist relations because extraction is premised upon using up its resources (in contrast to, for example, agriculture, which, depending on the practices followed, may be destructive to the land, but is usually performed with the intent of maintaining the land for future use). Furthermore, resource extraction has been, historically and contemporaneously, predominantly organized through unequal global power relations and the ideologies and political economies of colonialism.

Canada's extractive industry emerged as part of its settler colonial project and enabled that project. By contrast, modern resource extraction in Africa – especially diamond mining – was established as part of British imperialism. Indeed, the beginnings of the modern diamond industry in South Africa are inseparable from Britain's imperial expansion into Africa. The De Beers charter gave the company the mandate to "take steps for good government of any territory, raise and maintain a standing army, and undertake warlike operations" (Smillie 2010: 35). The De Beers diamond operations expanded throughout Africa alongside British force, with Rhodes as a key figure, and in the context of the larger European scramble for Africa. Indeed, recall that the Eureka Diamond was discovered in 1867: from this date to 1914, Europe expanded its imperial control from less than one tenth of the African continent to more than nine-tenths (Chamberlain 2014). The diamond industry both flowed from and contributed to the coffers funding British activities in Africa. Thus, the violence of the modern diamond industry cannot be untangled from the violence of imperialism; as Galeano puts it, "the fanatical mission against the natives' heresy was mixed with the fever that New World treasures stirred in the conquering hosts." Extractive plunder was made possible and justified through the colonial racisms that targeted the populations of extractable spaces as workers and heathens.

However, it was not this long history of colonial dispossession that threatened the reputation of the diamond industry. No, the violence of the colonial extraction of diamonds – as with luxury minerals in general – was masked through the carefully constructed distance between sites of production and sites of consumption (Hartwick) and through the shared stake in, and normalization of, imperial and colonial relations by producers and consumers in the Global North. Thus, it was South–South violence (not North–South violence) that captured the attention of activists and consumers in the Global North. In the 1990s, two NGOs launched campaigns that succeeded in bringing conflict diamonds to global attention. As a result of those campaigns, concerns over "blood diamonds" were articulated broadly and swiftly across global forums, from *Blood Diamonds* (1996), to the James Bond film *Die Another Day* (2002), to, ultimately, the Kimberley Process, an innovative international regulatory body that shapes the diamond industry today.

Broadly speaking, conflict diamonds are rough diamonds pulled from alluvial deposits at unregulated mines and are tied to armed conflict (for the most part, intra-state). All diamonds are produced in kimberlite pipes; major diamond operations (like those in Canada) mine the diamonds directly out of the pipes themselves. This is the most intensive and expensive, though ultimately the most profitable, form of diamond mining. Alluvial diamonds were formed in kimberlite pipes but have been dispersed through erosion. Alluvial diamond deposits are superficial – often within a few feet of the earth's surface – and thus much easier to mine (Smillie 2010). Because they require far less capital, these deposits are sites of what is called artisanal or illegal mining, depending on the surrounding politics, and the sources for conflict diamonds.[5] Conflict diamonds were not a new phenomenon when they came to global attention in the 1990s. However, it was the scale of both the market and the human cost of conflict diamonds in Sierra Leone and Angola that distinguished this form of extractive violence. Grant (2012) writes that by the mid-1990s, "in Angola, UNITA [The National Union for the Total Independence of Angola] was earning as much as US$700 million per year from rough diamonds (Global Witness 1998; Cortright et al., 2000); and in Sierra Leone … at its peak, the RUF's [Revolutionary United Front] annual income from rough diamonds was estimated at US$125 million" (162). The violence associated with the conflicts funded by these diamond industries has been staggering: in recent decades, an estimated four million people have died in wars involving conflict diamonds (Bieri 2010). Regarding Sierra Leone, Smillie and colleagues (2000) write that

> diamonds have fueled a conflict that has destabilized the country for the better part of a decade. Seventy five thousand people – most of them civilians – have lost their lives. Rebel butchery has left thousands of women, men and children without hands and feet, disfigured physically and psychologically for life. At different times during the crisis, as many as half of Sierra Leone's population – more than the entire population of Kosovo – became displaced or were refugees. Schools, hospitals, government services and normal commerce ground to a halt in all but the largest urban centres. Mineral resources which should have been available for development were used to finance the war, robbing the potential beneficiaries and an entire generation of children, putting Sierra Leone very last on the UNDP Human Development Index. (9)

This excerpt was written by Partnership Africa Canada (PAC), one of two NGOs that released reports around the same time chronicling this horrific violence and the role the diamond industry played in fuelling the conflict. In 1998, Global Witness, an NGO based in the UK, released *A Rough Trade*, which pointed to De Beers's role in the Angolan conflict. They wrote, "Given that De Beers were, according to their own reports, buying a substantial proportion of Angolan rough diamonds, at a time when a large section of the country's diamond mines were under UNITA's control, one could conclude that the drive to keep the lucrative outside market was a primary concern – despite the consequences this might have for the people of Angola during this period" (3). In 2000, PAC published *The Heart of the Matter: Sierra Leone, Diamonds, and Human Security* (Smillie et al. 2000), which linked the conflict in the country to the diamond industry, highlighting the role of the international diamond market, including Belgium, where the majority of diamonds are sorted, and De Beers, the "largest and most powerful player in the diamond world" (Smillie 2010: 4). Both NGOs called attention to the roles of Western powers in these conflicts and demanded international action. This set in motion international coalition-building that would result in a remarkably quick response to "conflict diamonds."

National and international organizations were spurred to action by the NGOs' campaigns (Grant and Taylor 2004). The Kimberley Process is named after the South African town where the international coalition held its inaugural meeting; significantly, Kimberley is also the town where the first major modern diamond find was made, a place Smillie calls "the spiritual capital of the modern diamond industry" (178). The Kimberley Process, initially led by a number of southern African countries involved in the diamond industry, refers to the early intergovernmental meetings responding to conflict diamonds. The process gained international institutional weight through UN Resolution 55/56, passed in December 2000, which mandated an expanded Kimberley Process (392). The early and expanded forms of this process focused on an international diamond certification scheme. Two years and many meetings later, the Kimberley Process Certification Scheme was introduced, premised on participation by states in mobilizing their relevant laws, regulations, and policy instruments to monitor and regulate the diamond trade within and across their borders and to guarantee these activities through a national certificate. This means that recruiting states into the process has been a crucial element for regulatory success.

Anyone familiar with the tortoise-like pace of this sort of international politicking will likely be experiencing whiplash as a result of the speed with which international activism turned into multilateral policy-making turned into international policy. That said, scholars and interested parties have pointed to the shortcomings of the Kimberley Process and the ongoing challenges it faces. Grant and Taylor (2004) have noted the difficulties of tracking diamonds through the global commodity chain, including the lack of transparency in some governments' certification processes and the lack of participation by some diamond jewellers. More broadly, Smillie (2010) argues that the Kimberley Process lacks teeth – that is, the capacity to respond to non-compliance in a meaningful way. However, as Grant (2010) rightly notes, shortcomings aside, the Kimberley Process has made it more difficult to trade in conflict diamonds – a feat of no small importance, given the devastating violence that initially spurred the conflict diamond movement.

What is arguably lacking in the policies and programs responding to conflict diamonds is not international governance acumen (indeed, the efficacy of such transnational multilateral change-making is truly remarkable); rather, the gap in the response to conflict diamonds is located in the space between conflict and "peaceful" resource extraction. The conflict diamond campaign focused on extraction in African countries by African political bodies, yet the broader history of extraction is largely about the violence of extraction by the Global North in the Global South. However, even while the stain of blood has threatened the diamond industry, centuries of violent colonial extraction have not posed a significant threat to the branding of luxury minerals; instead, these colonial relations continue to reappear in new forms, as we will see in the chapters that follow. And indeed, while De Beers faced significant criticism throughout the time of the conflict diamond "scandal" (as though it were a temporally isolated incident), the diamond giant survived by reaffirming its role as a legitimate actor in the global market. It is no wonder that De Beers eagerly participated in the new certification processes and forged partnerships with NGOs critical of conflict diamonds. In the end, it was the small-scale African miners who became the "illicit" actors.

Le Billon writes that conflict diamond campaigns succeeded in part because they combined the sexy imagery of diamonds with "the repulsive fascination of Africa in the occidental imaginary" (Amselle 200 in Le Billon 2006: 788). He argues that "in many conflict diamond narratives, political leaders, combatants and miners were often indiscriminately conflated into one single category of 'greedy thugs'; even if the status of alluvial diamond diggers as victims and legitimate political agents rightly seeking social change, as well as their social diversity have been documented" (789). Conflict diamonds were framed not as a harrowing but explicable expression of local articulations of the global processes of accumulation, extraction, and colonialism, but rather as a regional aberration from the proper rules and regulations of the global political economy. Scholars wrote of the resource curse, here, not as an aspect of uneven development, but as a politics of greed (795). Thus, while the response to the immediate crisis was effective, some iterations of the conflict diamond narrative reproduced racialized hierarchies that had long been disseminated, and justified, by centuries of colonial extraction.

On this uneven racial landscape, Canada emerged as the "white" saviour of the diamond industry. Canada has played a prominent role throughout the response to conflict diamonds. In 1999, just as the NGO-led exposés of conflict diamonds were breaking, Canada was elected to the Security Council and asked to chair the Iraq Sanctions Committee. Robert Fowler, Canada's Ambassador to the UN, chose instead to focus on Angola, and he led an investigation into Angolan diamonds. The result was the Fowler Report, released one year later, which named and shamed those complicit in Angolan conflict diamonds (Smillie 2010: 178). As the regulatory approach developed, Canada would emerge as a strong proponent of the Kimberley Process (Santarossa 2004). As Grant (2012) notes, a major outcome of the Kimberley Process has been international capacity-building in diamond certification and regulation. Canada has been activist in this regard, going above and beyond its commitments. The Royal Canadian Mounted Police (RCMP), for example, has been developing various forms of diamond fingerprinting to track diamonds along chains of manufacturing and distribution (Smillie 2010: 181).

However, much more than its participation in regulatory schemes, Canada's own diamond mines have served as a "safety valve" for the international diamond industry. The same decade that brought widespread Western awareness and international action against conflict diamonds in Africa saw the "discovery" of diamonds in the Canadian Subarctic. What, then, has been the relationship between the establishment of a Canadian diamond industry and Canada's role in denouncing the "blood diamonds" mined elsewhere? Or, to put it another way, was Canadian participation in the Kimberley Process a crass capitalization on conflict (Duncan 2000 in Le Billon 2006), a convenient means of securing the global market for a Canadian good through processes of international governance? Speculation on this question speaks to the entangled imperatives of foreign policy, international development, and global capital. And while a fly-on-the-wall history might answer questions about what directives were flying between federal departments, or what phone calls were being made between Bay Street and Ottawa in the hurried years that brought us both the international policy response (the Kimberley Process) and the commodity response (Canadian diamonds) to conflict diamonds,[6] ultimately, the degree to which Canada's participation in the Kimberley Process was driven by economic rather than humanitarian concerns is somewhat beside the point. Just as the violence associated with conflict diamonds cannot be pinned on a few warlords, the success of the Canadian diamond industry cannot be attributed to backroom deals or cynical policy-making. Rather, the Canadian diamond industry – and its imagined exceptionality, especially vis-à-vis the Global South – emerges in continuity with the uneven racial landscape of global capital. Canada, a country built upon land and resource grabs, has recast its settler colonial past as a tale of rugged discovery and insisted upon its innocence, time and time again. This innocence, literally branded into the diamond mines, has launched Canadian diamonds down global commodity chains into the hands of the (newly "ethical") consumer.

III. Canadian Diamonds

The Kimberley Process's primary mode of regulating diamonds – that is, a diamond certification process guaranteed by state governments – is nothing without participation; as Grant and Taylor and Smillie note, this reliance has led to governance gaps in the post-Kimberley years. As one would expect, Canada is an eager participant in Kimberley-mandated regulation, and then some. The certification of Canadian diamonds is a multi-scalar process that unabashedly blurs the line between certification and branding. At a national level, Natural Resources Canada adheres to the Kimberley Process Certification Scheme through legislation that requires rough diamonds to hold a Kimberley Process Certificate before they can be exported; that certification process enables the tracking of international shipments of diamonds (NRCAN 2019). At the territorial level, the Kimberley Certification process is complemented by the more comprehensive Government Certified Canadian Diamond process. This territorially administered certification guarantees that a diamond was mined, cut, and polished in the NWT (GNWT 2019a). This secondary certification sought to encourage the NWT diamond manufacturing industry (cutting and polishing), thereby keeping value-added activities in the North. Keeping manufacturing in the NWT has been a goal since the beginning of the diamond mines, spearheaded by Stephen Kakfwi, a Dene activist who served as a minister in the territorial government. Kakfwi was well aware of the North's location in processes of uneven development, wherein resources were extracted from the periphery and manufactured in centres of global capital.[7]

Despite ongoing territorial efforts to subsidize the diamond manufacturing industry – an industry largely reliant on foreign workers, notably from Armenia (Stueck 2002) – the NWT diamond manufacturing industry has yet to flourish.[8] This is for a variety of reasons, including the time it takes to train as a diamond cutter and the ease with which diamonds can be sent to regions with established diamond cutting industries (traditionally Belgium, but the diamond cutting industry is now established around the globe, including in Israel and India). Thus, the GNWT certification of "fully" Canadian diamonds has had limited take-up in the industry. The GNWT has drawn more industry interest at the blurry boundary between certification and branding. The GNWT Diamond Policy (2019a) identifies the option of granting diamond manufacturers the GNWT trademark – a Polar Bear – without delineating clear location requirements for cutting and polishing. This option likely emerged as a response to the company-led "Canada" branding of diamonds (i.e., Polar Bears or Maple Leafs lasered into diamonds; Stueck 2002) and to the Canadamark certification, a guarantee underwritten not by government but by BHP Billiton, the original owner of Ekati, the NWT's first diamond mine. Canadamark diamonds are lasered with a tracking number and logo: consumers can then use the tracking number to ensure that their diamond is as Canadian as they are told. However, like the Kimberley certification, the Canadamark guarantees only the diamond's location of origin (Schlosser 2013: 169); no environmental or labour standards or fair trade provisions come attached to this certification. As Schlosser writes, "such schemes potentially conflate the security or transparency of a commodity chain with the ethics of it" (169).

And indeed, while the certification of Canadian diamonds amounts to a set of overlapping, ever-shifting, ever-multiplying processes, in the end the only guarantee is the location of origin. The Canadamark website asserts that "all Canadamark diamonds are responsibly mined in Canada's Northwest Territories with respect for the sustainability of the natural environment and people living in the community," but this guarantee, as with all Canadian diamond certification, relies upon trust in the Canadian state's social, labour, and environmental regulations and monitoring. In this way, Canada's political brand and the diamonds' commodity branding reinforce each other. The Canadian state has long punched above its (economic) weight in international relations, in part because of its special relationships with the imperial powers of the day, but also because of the normative power associated with Canada's upstanding global reputation. Certainly, its reputation for social responsibility domestically and internationally has shifted from one government to the next. Prime Minster Stephen Harper, for example, was unapologetically direct in tying Canadian international development funding to Canada's economic interests (Butler 2015), and his government came under increasing international scrutiny for its treatment of Indigenous peoples (Amnesty International 2014). In 2015, the government of the next prime minister, Justin Trudeau, rode in on a wave of promises to change Canada's domestic and foreign policy. Two key planks to his election platform had been to address the Syrian refugee crisis and build respectful nation-to-nation relationships

with Canada's Indigenous peoples. But in the years since, it has become painfully apparent that Trudeau is willing to pursue an agenda
of social change only insofar as it will not interrupt the daily workings
of global capital or perturb those whose interests these daily workings
serve. And it is this marriage of Canada's "open for business" attitude
with a nice guy façade, as exemplified by the Trudeau government, that
has made the country so appealing to the diamond industry and similar businesses, which can use Canada's respectability to hold up their
"ethical" brand without worrying about interruptions in the workings
of extraction.

In this way, diamond mines rely on what Schlosser calls "nation-
branding," privileging "an abstract, universalized notion of Canadian
nationhood" (2013: 53) that combines Canada's reputation for responsibility with striking imagery and narratives of a pure, empty land. Indeed,
in promotional materials, Canadian diamonds often appear emerging,
as if by magic, out of the Arctic landscape. Schlosser identifies the diamond industry's blunt branding techniques as "commodity racism,"
in that they implicitly or explicitly compare diamonds from the "great
white (in all ways) north" to imagined impure diamonds: "Mobilizing
ethical consumption discourse in this way exemplifies the use of space
to increase exchange value, and how that exchange value reflects the racialized transnational identities from which it evolves" (175). The purity
of Canadian diamonds is rooted primarily in the mines' location, and
meanwhile, both the Canadian state and the diamond industry shore
up their image of responsibility through commitments to Indigenous
participation, the subject of the next chapter. In this formulation, em-
ploying northern Indigenous people at the diamond mines is framed as
a project of benevolent modernization. That these two hallmarks of responsibility – the untouched beauty of the North and the modernization
of the North – are fundamentally at odds with each other does not seem
to pose a problem to industry branding; instead, the Canadian brand
is embraced as both beautifully natural and reassuringly modern –
as offering the "new diamond" to the concerned consumer.

However, while the Canadian diamond brand is awash with perceptions of novelty and exceptionality, extractives in Canada are, of
course, anything but novel or exceptional. It was the fur trade that first
brought settlers to Canada, and the extractive industry has been the
engine of the Canadian economy ever since (Innis 1927; Watkins 1963),
expanding into, for example, lumber exports and, by the nineteenth
century, mineral mines. While the mid-twentieth century saw increased
state efforts to diversify the Canadian economy, extraction continued to
be a mainstay, with internal colonial resource extraction increasingly

accompanied by Canadian government and business activities in international resource extraction. International extractive activities line Bay Street coffers and are inscribed deep into Canada's foreign policy and international development practices: Canada has signed nearly two dozen Foreign Investment Protection Agreements (FIPAs) with countries in the Global South in order to guarantee its extractive activities (Gordon and Webber 2008: 66), and, as a result of Harper's development agenda, much of Canada's "development" money is tied to corporate social responsibility (CSR) for international mining operations and community development in mining regions (Butler 2015). Canadian mining companies have dispossessed communities and enacted violence around the globe; not surprisingly, the commodities emerging from these mines are not imbued with Maple Leafs or Polar Bears.

The Canadian brand associated with diamonds thus stands in sharp contrast to this country's record as a direct foreign investor in extractive projects in the Global South, notwithstanding some publicity for Canadian social development projects linked to Canadian mines.[9] FIPAs tend to shield Canadian mining companies from environmental and labour regulations (both Canadian regulations and the regulations of the country hosting the mine), and the geographic and cultural space between Canada and its international sites of extraction is used to foster impunity from public or political scrutiny. While Canada's domestic extractive legislation remains exceptionally permissive (Hoogeven 2016), operations of mines internal to Canada attract much closer scrutiny. Canada's domestic and international extractive operations are distinguished by a "responsibility paradox," wherein Canada attempts to reconcile its aggressive pursuit of extractive capital with its liberal political legitimacy. Internationally, efforts to resolve this paradox largely take the form of development funding attached to extractive projects, rather than any changes made to the relations of extraction themselves. At the same time, social movements and public pressures of the past forty years have demanded changes in how resource extraction is done within Canada, and attempts at "responsible development" have infiltrated extractive operations. The contradictions inherent in this contested and limited shift play out against the settler colonial backdrop of the mixed economy.

Conclusion

The relationship between diamonds and the "love industry" intensifies the need, common in luxury commodities, for diamonds to be distanced from the commodity chains from whence they came. As a result of sustained rhetorical labours on the part of the diamond industry, diamonds in boxes and on fingers are attached to hetero-monogamous and class aspirations rather than to the colonial extractive operations that supplied them. In the late twentieth century, the intensity of the violence attached to small-scale diamond mines – which were funding conflicts in Sierra Leone and Angola – shook public perceptions of the industry. Lax regulations came to light, and the complicity of big diamond players (including De Beers and the Belgian government) was revealed by small NGOs to huge effect. The result was the Kimberley Process Certification Scheme, which compelled countries with diamond mines to monitor and regulate their industry, tracking imports and exports.

The Canadian diamond industry emerged around the same time and came to represent an ethical alternative to "blood diamonds." Canada has enthusiastically participated in the Kimberley Process, first in international advocacy and diplomacy efforts, and more recently in international capacity-building efforts and in its own regulation process. Indeed, Canadian diamonds more than meet the Kimberley Process certification standards, and the exceptionality of Canadian diamonds (worry-free, high-quality gems) is a major part of the industry's branding, literally etched on the diamonds themselves, in the form of Polar Bears or Maple Leafs. Canadian diamonds have been made exceptional, not only to (potentially conflict) diamonds from elsewhere, but also to Canada's international extractive operations, which are a major driver of Canada's economy but are not included in the Canada brand. In the following chapter, I examine the NWT diamond-mining regime, exploring that which is new alongside the continuities with past forms of settler colonial extraction in the northern mixed economy.

5 The NWT Diamond-Mining Regime

Well, imagine. Our economy is based on something nobody needs.
                                                    – Talking Circle One

Canada is a political economy built on resource extraction, and the role of resource extraction in the settler colonial "making" of Canada is elevated in the North. Yet against the backdrop of so many booms and busts, the Canadian diamond industry had the audacity to claim novelty. There is, indeed, something new about this extractive regime, though it is not the sheen of imagined "purity" to which the industry professes. Rather, following contemporary trends in approaches to extraction (Pini and Mayes 2014), the diamond mines operate through a FIFO structure, wherein workers live at a distance from the mine and are flown in for shifts. Consequently, the FIFO diamond-mining regime (and by regime, I mean the relations of production engaged in the extractive process and the political and economic institutional apparatuses that shape these relations) effectively emphasizes variable capital (labour) over fixed capital (infrastructure). Furthermore, the regime enacts a new tempo to labour that takes shape through a spatial articulation of capital's separation of (feminized) social reproduction from (masculinized) capitalist production. This labour restructuring, as articulated in the departments of capitalist production, social reproduction, and subsistence production, is the subject of the three chapters of Part III.

In this chapter, I discuss the establishment of the NWT diamond industry and the ways in which the diamond-mining regime has been shaped by northern Indigenous social movements and the Canadian state's (mis)interpretations of their calls for self-determination. I argue that the diamond-mining regime represents an approach to Indigenous communities and people that is distinct from that of its predecessor, the gold mines. It is an approach oriented toward participation *by* Indigenous communities, if not any sort of fulsome responsibility *to* Indigenous communities. Instead, Indigenous rights are reframed as the "right" to participate in diamond mining. Thus, while the Canadian diamond industry distinguishes itself as "responsible" luxury, colonial dispossession persists alongside a neoliberal move toward Indigenous incorporation. In this way, at the same time that the diamond-mining regime introduces new elements in its approach to extraction, it is also another iteration of something not at all novel. Diamond mining is one in a long line of northern settler extractive projects, from the fur trade to the Klondike Gold Rush to the uranium mines of the mid-twentieth century.

The analysis in this chapter unfolds in two sections. In Section I, I introduce diamond mining against the backdrop of the gold mining industry and the turbulent decade – the 1990s – that saw the end of the gold-mining era and the beginning of diamond mining. The gold mines instilled a "temporary permanence" in the settler experience of the mixed economy, one ultimately characterized by the rupture inherent in boom-and-bust economies, culminating in a violent end to the gold-mining regime that played out alongside the "discovery" of diamonds. In Section II, I situate the diamond-mining regime's new approach to northern Indigenous communities in the history of northern Indigenous organizing. The diamond mines are a site of tension between place-based Indigenous demands for recognition and temporal capitalist imperatives for accumulation. In analysing the two primary NWT policy instruments related to mining/community relations – socio-economic agreements (SEAs) and impact benefit agreements (IBAs) – I argue that these are rhetorical attempts to resolve this tension. This is, however, a rhetorical resolution only, as the diamond-mining regime's attention to Indigenous participation is limited by a conception of Indigenous identity rooted in liberal capitalist individualistic rights-based structures.

I. Boom and Bust and Back Again: The Establishment of the Diamond Mines

a) Boom and Bust in the Mixed Economy

The previous chapter emphasized the violent continuities that tie extractive projects across place and time. The uneven, cyclical nature of economies dependent on the extraction of raw materials from the earth is a well-trodden subject. It has been explored by world systems theorists (Kohl and Farthing 2012), who link uneven global development with the extraction of raw materials from the "periphery" for commodity production in the "core," and as the "resource curse" (see, for example, Karl 1999; Le Billon 2001), in a diverse body of literature seeking to understand the relationship between conflict and extraction (as with conflict diamonds). Bellamy Foster and Clark (2004) urge us to approach the instability associated with extraction, not as an aberration from other forms of capitalist production, but rather as an expression of the instability and violence of ecological imperialism and capitalism itself. The following analysis of boom-and-bust dynamics follows that imperative. Framing the boom and bust of northern extractive economies through the Dene concept of rupture, I characterize the place-based experience of the breakdown of one extractive regime and the subsequent development of another as a series of multilayered ruptures: the immediate rupture of a boom-and-bust regional political economy (here, the bust of the Yellowknife gold mines that preceded the diamond boom) is set against the backdrop of the *ongoing* and shifting tension in the relationship between subsistence and capitalist production in the northern mixed economy.

The development of the FIFO diamond-mining extractive regime occurred in the context of a crisis with its predecessor, the gold mines. A successful strike at Giant Mine in 1980 and the proliferation of high-paying government jobs in Yellowknife had bolstered the high-wage model of mine employment at both Giant Mine and Con Mine, discussed in chapter 3. However, gold prices fell in 1981 and 1982, and, over the following years, the mines became increasingly run-down (Selleck and Thompson 1997: 9). Giant Mine, in particular, struggled to turn a profit, and in 1990 it was sold to Royal Oaks, a small company that implemented a lean, neoliberal approach to operations. Selleck and Thompson write that the Royal Oaks business model was "to buy mines with high operating costs when the price of gold is low, slash costs, survive until the price of gold rises, rake in profits and shareholder praise, and then buy something bigger" (1997: 13). The new owners and their management style had enormous implications for the town of Yellowknife. Almost immediately, long-time supervisors were let go and labour conditions changed dramatically. The new model, an expression of capitalism's drive to escalate the exploitation of human labour, introduced an intensified temporariness to employment that foreshadowed the diamond-mines' approach to extraction that would follow. In employer/union bargaining, which occurred less than a year after the change in ownership, the employer's attempts to roll back twenty years of membership gains were met with staunch refusal by the activist union. A seven-month strike ensued, a time that is remembered with sadness by residents of Yellowknife (Interviews 105, 115). The employer immediately initiated strike-breaking tactics, including hiring replacement workers. The replacement workers were often unemployed people, flown from Ontario into violent working conditions: they worked and slept at the mine for a month and then were flown back to their homes for two weeks off (Selleck and Thompson 1997). The workers slept at the mine because, at this point, employer/employee relations had turned violent and it was too dangerous for them to leave the mine site and enter Yellowknife. This was a bitter and often-forgotten first taste of FIFO work in Yellowknife.

In a small town that relied on mining jobs, the employer tactics had devastating consequences. Over the course of the strike, violence escalated: the employer brought in private security and repeatedly petitioned the RCMP and the Department of National Defence (DND) to use force against the strikers. Some of the strikers resorted to sabotage, including setting off small bombs to halt production and intimidate the replacement workers. On 18 September 1992, a planted bomb exploded in a mineshaft, killing nine workers. The deaths of the workers and the murder trial continue to be a source of tension and pain in Yellowknife.[1] Heather, who lived in Yellowknife at the time, but worked outside the extractive industry, described town life during and after the strike:

> I heard the level of bitterness. And the stuff, the level of hatred, it was absolutely, I don't know how this town got to be the way it was. And after that bomb went off that day, there was an eerie fog that came down that night ... because we couldn't, and you couldn't go uptown that whole year because they closed, there were people who would actually go out scab hunting ... And I don't think that we knew how to heal properly in those days. We really didn't. (Interview 105)

Jenny, a young woman who had worked for the diamond mines, said that while the breakdown of the gold mines is recent history, it is rarely discussed in Yellowknife today:

> There's this chunk of history that exists that not a lot of people really like to talk about, but it was pre-diamond mine that Yellowknife was a wild, unlawful, depressive place. And that came out of a strike and murder situation, but you know a lot of the economy tanked. People lost jobs. There were a lot of hard feelings. And you kind of get this picture of what the

post-gold, pre-diamond picture would have been like, and it's scary ...
Where a lot of people have very dark memories. And all of a sudden, that
became a memory. And that feeling just doesn't exist. It was very quickly
forgotten. (Interview 115)

Heather's and Jenny's articulations of Yellowknife's recent history
speak to the temporary nature of the boom-and-bust cycle and to the
way it imposes a social dependence on resource extraction. The 1990s
began with intensified levels of capitalist exploitation, imposed by an
extractive company that built its profits by paring down operations to
a veritable shell, that led to violence. The wound of unresolved conflict
was paired with the financial insecurity associated with an industry on
its way out. The fear and desperation that settlers felt as they saw their
source of employment vanish in front of their eyes is a reminder that
one cannot dismiss the clamour for resources as a movement emanat-
ing solely from elite interests. Residents of the NWT mixed economy,
especially those living around Yellowknife, a town that grew out of the
gold mines, had built their lives around an industry (some, over generations) that was disappearing rapidly before their eyes. They were
faced with the prospect of meeting the high costs of their day-to-day
social reproduction without the steady work they had come to expect.

However, while for many settlers the decline of the gold mines was
the end of a time of abundance, for the Tłı̨chǫ and especially the Yellowknives Dene, the gold mines had been a violent imposition on their
land and mode of life. This imposition would not soon be forgotten.
Jenny put it this way: "Now you have 237,000 tonnes of Arsenic Triox-
ide that people carry as the scar" (Interview 115). Over the lifespan of
the gold mines, northern Indigenous peoples had organized and made
significant gains that would shape the FIFO diamond-mining regime.
As the gold mines closed and the possibility of diamond mines materi-
alized, it was clear that neither private industry nor the Canadian state
would be able to develop an extractive regime without Indigenous
consultation and participation. Thus, the development of the diamond
mines was shaped by the contradictory influences of newly austere and
temporary labour conditions, in line with the changing labour practices common to neoliberal restructuring, and the political imperative
to consult Indigenous communities and recruit Indigenous workers.

Diamonds were officially discovered in the NWT in 1991. I write
"officially discovered" because it was common local understanding
that the Dene had "discovered" diamonds but had chosen not to mine
them because of the difficulty involved and their lack of utility. Chuck
Fipke, a prospector, worked with Inuit traditional knowledge holders[2]

in his search for diamonds and found a deposit of them around 300 kilometres northeast of Yellowknife. The deposit held diamonds of a quality and quantity worth pursuing, so in order to develop the extractive site, Fipke partnered with Broken Hill Proprietary Company (BHP), an Australian resource giant with more than 40,000 employees in twenty-five countries. On 14 October 1998, after a fast-forwarded consultation process, BHP opened Ekati Diamond Mines; now 80 per cent owned by Dominion Diamond Corporation, a Canadian diamond-mining company, it is newly named the Dominion Diamond Corporation Ekati Mine. Diavik Diamond Mine, just southeast of Ekati and the larger diamond producer of the two, opened in 2003. Diavik is 60 per cent owned by Rio Tinto, an international mining company based in Australia and the UK and operating in more than forty countries, and 40 per cent owned by Dominion Diamond Corporation. Snap Lake Mine, owned by De Beers Canada, was the third diamond mine in the region, as well as De Beers's first diamond mine outside of Africa. Opened on 25 July 2008, south of both Diavik and Ekati, it was the first entirely underground diamond mine in Canada (NWT and Nunavut Chamber of Mines 2008: 7). However, Snap Lake closed abruptly in December 2015 owing to insufficient profits, laying off 434 employees (Gilbert 2015). Gatcho Kué, which began its commercial operations in 2017, is the newest diamond mine. It is a joint venture between De Beers Canada and Mountain Province Diamonds. The mine is projected to have a twelve-year life and provides approximately 530 jobs a year (De Beers Group 2018).

b) Diamond Profit Flows

While Canadian political economy literature has debated what a reliance on resource extraction means for the strength and autonomy of the Canadian economy, the concern for some northerners is that the North is simply a depository of resources for the benefit of the southern Canadian metropole (as some would argue, an internal colony).[3] Rae (1976), for example, argues that the "monopolistic, large-scale, and externally directed character of the businesses which have shaped the economic history" of the North is paralleled by southern-driven governance of the area, which has "shared these characteristics to a remarkable extent" (75). As we have seen, northern settler administration did not arrive in the territory until the mid-twentieth century, and even then, much governance remained in the hands of the federal government, as compared to powers afforded to provinces. Regarding northern resources and their real and potential extractive rents, the federal

Image 3. NWT Diamond Mines (GNWT 2018a).[4]

government received all of the revenues generated by royalties on the diamond mines until 2014. Unlike provincial governments, the federal government maintained territorial subsurface rights, and thus, for all extractive projects – diamond mines included – the federal government set the terms with businesses and collected the resource royalties. Under this arrangement, the GNWT received transfer payments from the federal government. The only revenues they received directly from the mines were through fuel and property taxes. The GNWT estimates that since 2000 and up until 2014, "the three mines have collectively paid over $89 million in fuel taxes and over $165 million in property taxes"

(2014: 9). Compare this to approximately $1.6 billion generated *annually* in export profits by the mines (Government of Canada 2014).[5] Thus, from the end of the Second World War onward, the federal government's approach to the North has been driven by its keen fiduciary interest in extraction.

Given the material imperatives for extraction-friendly governance, it is no wonder that Ellen Bielawski (2003) characterizes the federal government as "the two-headed beast with conflicting mandates." The "second head" in Bielawski's metaphor is the federal government's role in Indigenous governance. While territorial and, increasingly, Indigenous governments are largely responsible for the provision of social services in the territory, because of the federal government's responsibility for Indigenous/state relations and the territory's high Indigenous population, the federal government, again, has a much heavier hand in sociopolitical governance in the territory, as compared to the provinces (notably, and discussed in more detail below, Indigenous governments negotiate with the federal government, not the territory, on land claim agreements). Thus, while the tension between federal commitments to supporting Indigenous well-being and settlement of land claims, on the one hand, and Canada's "free-entry" approach to extraction, on the other, is one that stretches across the country, it is found with particular intensity in the NWT.

What has changed since 2014 is not the degree of this tension, but the relative powers in the governance scales through which it is articulated. As a result of a devolution process, the GNWT now keeps up to 50 per cent of royalties generated through resource development, with the remainder continuing to go to the federal government. Of the royalties the GNWT receives, 25 per cent go to the Indigenous governments that were signatory to the Devolution Agreement.[6] Thus, in 2017–18 alone, the GNWT collected $66 million in mineral royalties, the vast majority from diamond mines (GNWT 2019c), on top of the taxes it had collected previously. While some research participants – most notably, those in government or private positions related to the administration of the mines – argued that devolution was an important move toward northern autonomy from federal control, others expressed concern that the devolution agreement would tie territorial politics more tightly to resource development, making a northern post-extractive future more difficult to pursue. And indeed, as evidenced by the *NWT Mineral Development Strategy* (GNWT and Nunavut Chamber of Mines 2014), a cross-departmental policy document researched and written in collaboration with the North's primary mining lobby group, mining is the cornerstone of the current territorial government's economic development plan. Writing fourteen years before the devolution of resource revenues, DiFrancesco (2000) made the important point that devolution is occurring in a climate of federal fiscal constraint, thus pushing the GNWT toward reliance on mineral development to fill the gaps left by decreased federal transfer payments. Extraction as *the* mode of northern development continues to characterize the upper echelons of territorial politics and is expressed in the territory's Mineral Resources Act, tabled in 2019 as a stand-alone act to govern NWT resource extraction.[7] Indeed, at the time of writing, as the diamond industry begins to contract, many territorial politicians continue to pin their economic hopes on initiatives for prolonging the life of existing mines and on junior mining ventures that could soften the forthcoming diamond bust.

Concerns about the post-diamond future are prescient, given the short life of the diamond mines. Ekati and Diavik generate more revenues and employ significantly more people than Snap Lake did or that Gatcho Kué has the capacity for, and both Ekati and Diavik are well past the halfway point in their economic life.[8] Ekati continues to pursue initiatives that it hopes will prolong its life, while Diavik has opted for planned closure: it is planned for 2025, with significant layoffs beginning in 2022 – a fairly unique approach in the mining industry, given that mines tend to operate until they can't, resulting in painfully abrupt job termination. While Gatcho Kué was marketed by Mountain Province Diamonds as the largest diamond mine under construction in the world, according to the GNWT, its small workforce will not offset job losses in the larger mines (14). As such, a common theme that was not solicited, but emerged, from interviews and talking circles was the temporariness of the FIFO diamond-mining regime. Only twenty years after the first diamond mine opened and subsequent socio-economic changes flooded the region, northerners must already look to the inevitable rupture of mine closure. Some participants discussed ways of extending the existing mines or developing new mines; others articulated ways of creating a post-extractive economy. But all shared concern about the potential violence of yet another socio-economic rupture. Many also articulated a curiosity about what alternative development possibilities might fill the space left by the fleeting, yet dominating, presence of the diamond mines.

For workers, as we will see in chapter 6, the *temporariness* of the diamond mines is experienced through the two-week in/two-week out *tempo* of the FIFO regime. That tempo, and the temporariness of the diamond-mining regime, sit in tension with the day-to-day and intergenerational social reproduction of a place-based orientation in the mixed economy. This tension is shaped by the diamond-mining regime's new approach to Indigenous relations, to which I now turn.

II. Diamond Mine/Indigenous Relations

In its relationship to northern Indigenous people and their labour, the diamond-mining regime represents, at once, a continuity and a discontinuity with past forms of extraction. As another iteration of resource extraction driven by the imperatives of private capital and the Canadian state (Parlee 2015), the diamond-mining regime sustains the tension between subsistence-oriented production and the extraction of surplus value in the mixed economy. However, the diamond mines also represent something novel, as northern Indigenous people have been newly targeted as potential mine workers and auxiliary business operators. The diamond-mining regime's new approach to northern Indigenous engagement and employment is shaped by the contradictory relations between Indigenous organizing for place-based rights, state efforts to maintain social legitimacy while asserting sovereignty over Indigenous lands, and the ongoing pursuit of new sites of extractive accumulation.

a) Indigenous Activism Takes On the State

Indigenous resistance to settler colonialism, so often obscured in historical narratives and dominant cultural touchstones, has shaped contemporary Canadian social relations. Thirty years ago, Abele and Stasiulus (1989) noted the decided gap in literature that acknowledged and sought to understand how Indigenous governance, resistance, and modes of life have influenced Canada's political economy. The influence of Indigenous activism is crucial to understanding the development of new modes of extraction in the North. In discussing the ways in which Indigenous activism has shaped the new northern extractive regime, I approach the relationship between Indigenous organizations (activist, community, government) and the Canadian state (at its multiple scales) dialectically, with the aim of honouring the hard-won gains of Indigenous organizations while recognizing the limitations imposed by settler laws, policies, and practices. In particular, I am attentive to the appropriation of Indigenous language of rights and recognition by the state and private capital, language that has been reinterpreted to shore up neoliberal approaches to Indigenous labour integration into northern extractive regimes.

The late 1960s and 1970s saw a new form of Indigenous organizing on the rise in the NWT (Watkins 1977, Coates and Powell 1989, Dickerson 1992, Abele 2009a, Coulthard 2014), characterized by a series of organizations, initiatives, and networks that challenged settler colonial governance and extractive processes and established the contemporary

Indigenous governance regimes that are shaping today's northern political economy. This period is sometimes described as the *beginning* of northern Indigenous organizing; however, that characterization denies the long-standing organizational forms of northern Indigenous peoples. Instead, it is more accurate to describe this time as the beginning of Indigenous organizing that directly targeted the Canadian state by using language and organizational forms that the Canadian state recognized (Coates 1985). The relatively laissez-faire state approach to northern governance up until the 1940s (Abele 2009b) gave northern Indigenous peoples less reason to politically organize *contra* Canadian state penetration. However, mid-twentieth century social interventions by the Canadian state coupled with intensified interest in northern resources – most notably in the form of the proposed Mackenzie Valley Pipeline – sparked new coalitions and new organizing strategies. For example, an unintended consequence of the residential schools' attempts to separate children from their families and communities was that large numbers of Indigenous children were brought together in these schools and taught the language of the colonizers (the master's tools). Some of the relationships carved out in these oppressive institutions would lead to radical decolonizing collaborations (Kakfwi 1977).

As a result of the intensified settler colonial intrusions by the Canadian state and private capital throughout the twentieth century, many northern Indigenous peoples saw the need to articulate a shared identity and common demands. Between 1970 and 1973, Inuvialuit of the Mackenzie Delta organized the Committee for Original People's Entitlement (COPE); the Inuit of Quebec, Labrador, and the Northwest Territories organized the Inuit Tapirisat of Canada (ITC), an umbrella organization for Canadian Inuit; The NWT Métis founded the Métis Association; and the NWT Dene organized the Indian Brotherhood of the Northwest Territories, an association that would develop into the Dene Nation (Dickerson 1992: 101–4). Coates and Powell (1989) write:

> In 1978, at a Fort Franklin meeting, it was decided to adopt a new name for the association [of Dene communities]. Stella Mendo, a Fort Norman delegate, caught the sense of the meeting: "When we name our children we give them powerful names. Names that are strong. Our land is strong. Our people are strong and our people are one. We need a strong, powerful name to tell the world who we are. Let us call ourselves the Dene Nation." (109)

At the same time that Indigenous communities were organizing toward self-determination, Justice Thomas Berger was commissioned to conduct an inquiry into the possible impacts of building a pipeline that would transport natural gas through the Mackenzie Valley. Rather than a quiet exercise in rubber-stamping, as might have been expected, the inquiry was a landmark study of the impact of resource extraction on Indigenous communities, one that continues to shape extractive expectations and policy to this day. Berger travelled throughout the NWT, holding community meetings and interviews. Many of these communities were already well-organized against the pipeline, and they used this opportunity to propel their concerns onto the national stage. At a hearing in Fort McPherson in 1975, for example, Philip Blake declared:

> We are a nation. We have our own land, our own ways, and our civiliza-
> tion. We do not want to destroy you or your land. Please do not destroy
> us ... I am sure throughout your visits to native communities, Mr. Berger,
> that you have been shown much of the hospitality that is our tradition as a
> people. We have always tried to treat our guests well; it never occurred to
> us that our guests would one day claim that they owned our whole house.
> Yet that is exactly what is happening. (in Watkins 1977: 7)

Little (2007) notes that Indigenous women participated heavily in this process, though the Canadian state and media tended to recognize Indigenous men as the leaders. Phoebe Nahanni, for example, led a landmark mapping project documenting the history of Indigenous subsistence land use in the NWT, and, thus, their claims to land (Nahanni 1977: 23).

The result of this organizing and the Berger Inquiry was a ten-year moratorium on the pipeline and a new precedent for community consultation preceding extractive projects. Justice Berger made note of the unceded status of Indigenous territory of the North and argued that no resource development should occur until land claims had been settled. The Berger Inquiry took place at a fortuitous moment nationally: Indigenous peoples had won the federal right to vote in 1960, and the 1969 White Paper and its assimilationist goals had been met with resistance and demands for a new federal approach to Indigenous governance.[9] As Dickerson (1992) notes, "the federal government was facing a growing number of claims throughout Canada, and, after problems with the White Paper, it was obviously necessary to clarify the legal basis for the claims" (106). A 1973 court decision, *Calder et al. vs. Attorney General of British Columbia*, involving the Nisga'a of the northwest coast, had further strengthened the precedent for recognizing Aboriginal title. This led to the recognition by then Department of Indian and Northern Development (DIAND) that Indigenous interests had not been adequately accounted for, and the 1974 creation of an "Office of Native Claims"

(Dickerson 1992: 106). It was in this context that the contemporary land claim negotiation process that continues to shape NWT politics and development began.

The late twentieth century saw the development of a number of northern land claims. In 1984, the Inuvialuit signed the Inuvialiut Final Agreement covering the northwest of the territory. In 1992, the Gwich'in signed an agreement with the federal government, followed by the Sahtu Dene and Métis in 1993 (McArthur 2009). In 1996, the agreement between Nunavut and the federal government was finalized and Nunavut became both a territory and the largest geographic area covered by a land claim. The Tłıchǫ Land Claims and Self-Government Agreement was signed in 2003, eighty-two years after Treaty 11 was signed by Chief Monfwi in 1921. The Tłıchǫ agreement is what is known as a comprehensive claim, in that it is a combined land claim and self-government agreement, the first of its kind in the NWT. The agreement divides the land claim area into four geographic regions. The Tłıchǫ have different rights in each: a traditional land use area; a resource management area (also part of traditional land use); land that the Tłıchǫ own in fee simple; and land that is designated as of "historical and cultural importance" to the Tłıchǫ but is not owned by them. The agreement allocates the following powers to the Tłıchǫ:

> Participation with a co-management structure to control permitting through an environmental management process, administer 39,000 km² (19% of their traditional territory) and manage health and social service delivery [the resource management area]. The Tłıchǫ own the harvesting right to the trees, forest and plants of the region, have the exclusive right to take and use waters flowing through the land and own the minerals under the land (Bill C-14). (in Gibson 2008: 240)

Yellowknife is within the geographic territory of the Tłıchǫ land claim. It was negotiated as exempt from the agreement (McArthur 2009: 202).

Behchokǫ̀, one hour north of Yellowknife, is the site of the Tłıchǫ Government, a body that has been growing since it was established. For many Tłıchǫ, the agreement, which led to self-government, has brought new hope. As John B. Zoe (2005) writes, through the agreement, "we have now got control of our own health and social services. We have designed our own healing path, and have ownership over our own land and businesses." At the same time, the land claims are a response to Indigenous calls for self-determination articulated through neoliberal, extractive, and Canadian state-building imperatives. Of particular note, while the diamond mines operate on traditional Tłıchǫ land, the subsurface rights granted through the land claim do not include the diamond mines.

As noted in chapter 3, the Yellowknives Dene, the community of Łutselk'e, and the Deninu Kų́ę́ of Fort Resolution have come together as the Akaitcho Nation. The Yellowknives Dene negotiate through the Akaitcho Nation and have yet to reach an agreement with the federal government. Their negotiations are tripartite: between the Akaitcho Nation, the NWT, and the federal government. In 2000, the Akaitcho signed a framework "to negotiate land, resource and governance issues without prejudice to Treaty 8 rights and obligations. One year later, the First Nation reached an agreement to protect lands on an interim basis, and in 2006 the Akaitcho and the GNWT reached a deal to temporarily protect land in and around Yellowknife while negotiations continue" (McArthur 2009: 203–5).

The very recent history of the gold mines and the development of Yellowknife on Akaitcho land (all without consultation or consent), the disastrous environmental legacy of the gold mines, and the ongoing socio-economic challenges the Akaitcho (specifically, the Yellowknives Dene) face as a community adjacent to the centre of political and economic settler activity in the territory have contributed to the resolve of the Akaitcho Nation to protect their treaty rights in these negotiations.

While most land claims in the NWT, unlike the Akaitcho claim, have now been signed, they are most accurately conceptualized as a piece of ongoing struggles for Indigenous self-determination, not the outcome of a completed process. Coulthard (2014) writes that, rather than being sites of organizing separate from subsistence, the early land claims proposed by the Dene Nation, and denied by the Canadian state, were informed by a place-based ethic, or grounded normativity, that critiqued colonial sovereignty and capital accumulation (64) and sought to protect subsistence production. The Canadian state's approach to land claims, conversely, is implicitly colonial capitalist in its approach to land-as-property and its assumption of the underlying sovereignty of the Canadian state. Indeed, it is no coincidence that land claims – which often involve protracted negotiations spanning decades – are suddenly greased with government motivation and new flexibility when extractive projects are on the horizon. Hayden King (2013) challenges the notion that modern and comprehensive land claims are something new, writing that they reflect instead "a very old phenomenon: that is, the marginalization and even dispossession of Indigenous peoples" (83). Similarly, Irlbacher-Fox and Mills (2009) characterize northern land claims as attempts at building the certainty required for extractive development:

> Canada seeks certainty through treaties – certainty of rights, of access to resources, of ownership of lands. Indigenous peoples seek recognition and a basis for achieving psychological, spiritual and material well-being. It seems that for Canada, treaties are an end, a final definition of a relationship, secured by the extinguishment of rights through certainty clauses. For indigenous peoples treaties mark a beginning, the beginning of a better life for beneficiaries and their communities and of a new relationship with Canada. (254)

Irlbacher-Fox and Mills are pointing to the strange marriage, in land claims, of the certainty – or aspired permanence – of land tenure desired by the Canadian state and private capital, and the commitment to recognizing distinct Indigenous relationships to land.[10] As Irlbacher-Fox (2009) notes, recognizing claims to land conceptualized as property is not the same as recognizing the right to self-determination, or self-government, nor is it the same as making space for alternative relations to land. Except for the Tłı̨chǫ land claim, the Canadian state has tended to separate recognition of self-government from land claims, a separation that, in and of itself, is antithetical to relational Indigenous ontologies of land (Coulthard 2010) and is potentially problematic for Indigenous communities' capacity for subsistence production in the mixed economy.[11]

Northern land claims, and their differing negotiations and implementation processes, are the subject of much contemporary northern scholarship (Dickerson 1992; DiFrancesco 2000; White 2002; Usher et al. 2003; Fitzpatrick 2007; White 2009; Abele 2009a; Irlbacher-Fox 2009; Laforce et al. 2009; McArthur 2009). For the most part, and with the notable exception of a portion of the Tłı̨chǫ land claim, the government approach to contemporary land claims has been to extinguish Indigenous claims to resource rights – in particular, to subsurface rights (i.e., rights to extractive resources). As Gibson (2008) explains, "land claims tend to be a central lever, and agreements tend to recognize that the lands have been occupied since time immemorial; however, federal control over mineral rights is affirmed. Precedence is always given to mineral rights, unless land claims are concluded and have provisions for protection" (126). As noted earlier, the diamond mines are on land that is now part of the Tłı̨chǫ land claim. However, at the time when agreements were negotiated between the first two diamond mines and Indigenous communities, no land claim had been settled. Federal jurisdiction over mineral rights, combined as it is with Canada's free-entry mining regime (Hoogeven 2016), whereby extractive companies can acquire rights simply (indeed, quite literally) by staking a claim, tends to relegate Indigenous communities to the role of "interested party" rather than rights-holders.

Thus, the diamond-mining regime developed through the regional norms that emerged through modern land claims processes, but because diamond mineral rights remained in the hands of the state, it was not primarily the particulars of the *specific* claims of Indigenous groups in the NWT that shaped the varying forms of commitments by diamond companies to Indigenous communities. Instead, it was the understanding – reinforced through decades of organizing, from grassroots community work to pan-territorial calls for self-determination, to the hard work of government-to-government negotiations – that northern resource extraction cannot occur without consultation with and participation of Indigenous communities. In contrast, the approach to Indigenous communities taken by the gold-mining regime had been one of "separate development." So, as a result of regional organizing bolstered by new national commitments to Indigenous rights, the diamond mines developed through an imperative for Indigenous consultation and participation. Under the land claims approach, Indigenous peoples were viewed primarily as "stakeholders" and a (dematerialized) cultural group, without acknowledgment of their distinct land-based ontologies and modes of production, and this enabled the diamond-mining regime to take an approach to "Indigeneity" as a floating identity-marker detached from land-based social relations.

b) (Mis)Interpretations and Indigenous Rights

In examining how Indigenous communities are approached by the diamond-mining regime, I focus on the dialectic between Indigenous calls for place-based rights and the state-building and capitalist imperatives to replace the groundedness of Indigenous relationships with a flexible – and absorbable – approach to Indigenous identity. The outcome of Indigenous organizing and the ensuing land claims and Indigenous participation in local, territorial, and federal northern governance has been an extractive regime that cannot ignore northern Indigenous communities. As a result of modern land claims, the Gwich'in, Sahtu, and Tłı̨chǫ are entitled to a percentage of resource revenue collected on public land in the Mackenzie Valley portion of the NWT, which is where the diamond mines are located. The Gwich'in and Sahtu are each entitled to receive annually 7.5 per cent of the first $2 million of resource revenues collected, or $150,000, and 1.5 per cent of any additional resource revenues collected, while the Tłı̨chǫ are entitled to 10.429 per cent of the first $2 million of resource revenues collected, or $208,580, and 2.086 per cent of any additional resource revenues collected (GNWT 2014a). As of 2016, these three First Nations had received approximately $40 million (NWT and Nunavut Chamber of Mines 2017). Additionally, the nine Indigenous organizations that were signatories to devolution share 25 per cent of the diamond royalties received by the GNWT, which, recall, splits the royalties 50/50 with the federal government. In the 2017–18 year, these Indigenous groups received a shared $8.4 million.

Given that the diamond operations are contributing more than $1 billion annually to the territorial GDP (in 2017, it was $1.8 billion) (GNWT 2018a), the royalties flowing to Indigenous organizations and governments are minimal, to say the least; though, certainly, the dollars flowing to Indigenous governments help greatly in funding governance activities and service provision at the community level. Indeed, when we analyse the impact that land claims negotiations have had on resource extraction projects – specifically, the diamond mines – a strange irony emerges: land claims processes have been resource-heavy legislative endeavours, yet ultimately, it is not land claims that provide the primary legislative or policy structure for the extractive approach taken by the diamond-mining companies. The land claims do ensure a base level of redistribution, but the agreements made between the diamond mines and the territorial government (SEAs), and the diamond mines and the Indigenous communities (IBAs) are the primary policy instruments shaping engagement between Indigenous communities and the diamond mines.

The SEAs and IBAs – institutionalized articulations of an attempt to rhetorically resolve the tensions between Indigenous modes of life and extraction – reflect a two-pronged approach to Indigenous/settler relations. As *communities*, Indigenous peoples are approached as stakeholders – that, is, as groups that should be consulted, compensated, and monitored for potential social "impact." As *individuals*, Indigenous people are approached as potential workers or as small business owners in auxiliary industries. These two approaches represent "being Indigenous" as a racialized identity separated from ideologies that may conflict with the requirements placed on the capitalist extractive worker. Dehistoricized and dematerialized in this way, "Indigenous" is a group or identity-marker of people who can be targeted, trained, and retained in the diamond-mining regime as workers or business operators. The settler colonial racialization of Indigenous peoples as groups tied to specific places (discussed in the Introduction and chapter 1) takes a new form here. While the relationship between Indigenous groups and the land explains the violent impulse toward genocide under settler colonialism (Wolfe 2006), in the contemporary political-economic context, the relationship between Indigenous groups and the land is reinterpreted as that of "stakeholder": a not-quite owner, in the sense of private property. Such a relationship is, however, compatible with the liberal capitalist mode of organizing land. In this way, Indigenous relationships to the land are redefined in terms amenable to Canadian state sovereignty and the structures of alienable and discrete private landholdings, which constitute the bedrock of capitalist accumulation. As such, the diamond-mining regime offers targeted opportunities for Indigenous individuals and Indigenous businesses to "succeed," but in terms constrained by capitalist imperatives. Significantly, this model has resulted in new access to high-wage labour for many northern Indigenous people; however, as I demonstrate in the third part of this book, the capacity to access this model of prosperity is highly unequal and structured in opposition to subsistence and social reproduction.

The SEAs are the territorial policy instrument for shaping and monitoring the impact of the mines on Indigenous communities. Up until devolution, the land on which the diamond mines were established was the jurisdiction of the Crown; as such, the GNWT could not legally establish terms for resource development (GNWT 2014). However, having extended the scope of environmental assessments (which are territorial jurisdiction) and the resulting agreements, the GNWT and the diamond companies developed the practice of establishing SEAs that encompassed corporate assessments of the potential socio-economic impact the mines might have on communities and the ways that diamond companies might mitigate harmful impacts while extending beneficial impacts (largely, employment). The SEAs have established a limited practice of monitoring the socio-economic impact of the diamond mines. The *Communities and Diamonds Report*, published annually by the GNWT, tracks community development through a designated set of quantitative socio-economic indicators, which are presented as related, though not exclusively so, to the diamond mines. While the indicators have changed over time (GNWT 2018a), they have tended to range from employment and housing, to community, family, and individual well-being, to cultural well-being and the traditional economy. Certainly, government attempts to recognize Indigenous modes of life and monitor and mitigate the community impact of extraction should be recognized as a step in the right direction; however, as Gibson (2008) pointedly argues, the quantitative model of assessment focuses on community deficits, approaching Indigenous communities as marginalized "Canadians" rather than as people tied together by a distinct set of historical and contemporary social relations of production. This obscures processes of settler colonialism. She writes that "the *Communities and Diamonds Report* seems bent on making people over in the image of settlers, adapting to market forces, improving housing status, and reducing the dysfunctions of 'transitions' that settler society deems inevitable in their narratives of change" (119).

In this way, Indigenous "poverty" is framed, not as a result of settler colonialism, but as a social problem to be fixed through employment and state benevolence (Adams 1995). Thus, extractive employment, and especially diamond mine employment, has emerged as the main development strategy employed by the territorial and federal governments in regard to NWT Indigenous communities. The SEAs map out specific targets for northern hires[12] and northern Indigenous hires. Employment targets have varied between mines and between operation phases; however, targets for northern hires have been at least 50 per cent, and the most recent northern targets for mining operations in Ekati and Diavik, the two longest-standing mines, were 62 and 66 per cent, respectively (GNWT 2018a). The target for northern Indigenous hires has consistently been 50 per cent – around 25 per cent of hires overall – which roughly matches their demographic representation in the territory. And indeed, throughout all operations, from 1996 to 2016 inclusive of all mine phases, northern Indigenous people have made up a cumulative 24 per cent of the mining workforce, with northern hires resting at 51 per cent (GNWT 2018a). The *Communities and Diamonds Report* communicates annual northern and northern Indigenous employment as it measures up to company targets. Newer iterations of these reports spotlight individual Indigenous training and employment stories (GNWT 2019a), building on the narrative of the possibility for individual success. Thus, while the community "stakeholder" approach to Indigenous/settler relations is somewhat limited, the individualized "participant" approach is emphasized in both the SEAs and the IBAs. Indeed, while IBAs vary, the promise of jobs is always their central component (Cameron and Levitan 2014), usually with accompanying company commitments to job training and development for the community in question.

SEAs are governed through the relationship between the GNWT and the diamond companies, whereas IBAs are bilateral agreements made between diamond companies and specific Indigenous communities. It is *individual* communities that make these agreements. The diamond companies insist that the IBAs be negotiated and implemented confidentially, making it nearly impossible for Indigenous communities to collaborate in negotiation or to learn from and build upon past negotiations by other communities. Alica, a young Indigenous woman and community organizer, characterized the effect of confidentiality in this way:

> I don't understand why these agreements are confidential. Why did we agree to confidential agreements when the other Dene groups, it basically made them all fight with each other? Because they couldn't tell each other what they were getting and all these rumours were going around, like, this group got way more than us. It caused a lot of fighting between groups and tension. And I felt it puts the First Nations group at a really big disadvantage when you're already going up against a huge international mining company. So I said, "I don't understand why we agreed to this." And the response I got was, "Well, to be honest, when they came in the nineties, we didn't have any experience with the mines or international corporations like that. We didn't have any experience with IBA agreements. So we were kind of like little kids up against these big guys and we just kind of went with it because we didn't know any better." And only now we're learning. (Interview 119)[13]

IBAs are not legislated, but they are expected practice: a new mine in the NWT would not begin operations without negotiating an IBA. That they are the primary mechanism for the flows of extractive "benefits" (undefined as they are) to Indigenous communities and that they are negotiated bilaterally (i.e., without government intervention) is, as Cameron and Levitan note, "a means by which the Crown's duty to consult and accommodate Indigenous peoples is privatized" (2014: 37). Drawing on Jamie Peck's analysis of state configurations under neoliberalism (2001), Cameron and Levitan argue that this is not a retrenchment of the state but rather a reorganization. Indeed, this is a reorganization that equates "responsible development" with employment and downloads that onus onto private extractive companies and the communities with which they negotiate.

More than the SEAs, IBAs have attracted international attention as an innovative approach to community consultation in the extractive industry. Indeed, Caine and Krogman (2010) argue that northern Canadian IBAs are an important historical development, representing an "opportunity for Aboriginal people to not only gain economically from resource extraction but also affect the trajectory and scale of development" (77).[14] IBAs provide socio-economic commitments to Indigenous communities and document local "buy-in," thereby giving diamond companies a greater chance at accessing a local workforce and enjoying political stability. Gibson (2008) writes that "from the mining company's perspective, the IBA represents decreased political risk. With communities on side, the risk of project failure or disruption through protest is eliminated" (129).

While employment targets are a core component of IBAs, IBA commitments demonstrate contingency and variability in the ways in which companies prioritize Indigenous participation in the mines. The first and most common way to involve Indigenous communities in the mines has been to employ Indigenous people in the mine labour force; however, throughout the northern diamond industry, Indigenous people and governments are increasingly being approached as potential entrepreneurs, business owners, and investors. James, a northern man in diamond management, noted that

> the Tłįchǫ will say that with Ekati, they were looking for jobs, and when Diavik came along they were looking for business. Business involvement. And then I think when Snap Lake came along, there were already businesses in places that Diavik helped support that were already demonstrating capability and knew the game and how to play ... So there's been a real step change there. (Interview 401)

These businesses are one of the major ways in which diamond revenue flows into Indigenous communities. In 2017, for example, the diamond industry spent around $324 million with northern Indigenous businesses. In total, since 1996, NWT diamond mines have spent around $6.452 billion (31% of total procurement) with northern Indigenous businesses and $14.587 billion with northern businesses in general (70% of total procurement) (GNWT 2019a). Compare, for example, the $324 million spent on Indigenous businesses in 2017 with the $8.4 million of resource royalties shared between nine Indigenous communities that same year and it becomes clear that the diamond extractive approach rewards a particular mode of "being Indigenous": the Indigenous entrepreneur/business owner and the Indigenous labourer, where labour is synonymous with extractive production. Indeed, the promotion of Indigenous business and the "Aboriginal entrepreneur" is far from unique to the diamond industry. The same strategy is promoted by scholars and policy-makers in the United States and Canada and taken up by both the US and Canadian governments.[15]

Emma, a white northern woman who had been involved with IBA implementation in Indigenous communities, echoed James and other research participants in noting that there has been an explosion of Indigenous-run businesses since the first diamond mine opened. However, unlike James, who characterized the new approach as an evolution from Indigenous workers to Indigenous business (the imagined capitalist trajectory from workers to owners of the means of production), Emma saw the approach to Indigenous communities and people as contingent rather than progressive:

> Training and employment; business; and financial payments: those are the three big parts of the IBA. So the business provisions have been scaled way, way back with Gatcho Kué. Before there were preferential contracting opportunities and evergreen contracts, which means that, for the life of the project, a specific contract was linked to a specific Aboriginal government. So, for example, at Diavik the catering contract goes to the Yellowknives Dene. No matter what happens … So evergreens are so past done … Because when the mines were first set up, you have to remember that there were, like, fifteen Aboriginal businesses. Now there's 250 or whatever there is. It's like an explosion of business. So part of it's reflecting that there was a time when you needed a lot of active inducement and now it's like, meh. It's already happening. The other part of it is the financial market and the viability of projects is completely different. So Ekati was like a slam dunk … Other projects like Snap Lake, the economics were always marginal. (Interview 402)

Emma's observations point to the conditionality of IBA commitments (conditionality based on the needs of the owner and the market at any given time). Indeed, because IBAs were confidential until recently, community expectations could not be set on a baseline established by previous negotiations. Instead, confidentiality allowed diamond-mining companies to negotiate community benefits and participation based on the market and the needs of the mine at the time of negotiation. Compare this enforced individualized approach to the community solidarity exhibited during the 1970s resistance to the Mackenzie Valley Pipeline. As such, research participants noted significant variation in IBA approaches to Indigenous relations and "community impact" over time, between the different mines, and between different communities. In the community workers' focus group, for example, participants agreed that availability of funds from the diamond mines for community projects – a feature of both SEAs and IBAs – was inconsistent and seemed to shift based on the profits of the mine and its public relations goals at the time.

The inconsistency in the FIFO diamond-mining regime's approach to Indigenous participation and community development is the result of the shifting temporal demands of capital and variability in *investment* in the labour force as a response to the demands of the market and the constraints faced by mine operators, rather than any sort of ambivalence toward Indigenous participation in and of itself. Indeed, prioritizing jobs for Indigenous people and contracting Indigenous businesses provides a number of benefits to diamond companies. Significantly, contracting Indigenous companies promotes the "responsible development" image of the diamond mines while also enabling national or transnational companies that are new to the North to rely on local expertise and reduce transportation costs. Perhaps most saliently, by implicating Indigenous communities in the management and investment sides of the extractive regime, diamond companies seek to reduce the likelihood of community-level resistance. James explained the benefit of this strategy to the industry with the following anecdote:

> When Diavik went forward, of course Diavik also had a socio-economic plan ... And of course they hire Th̩chǫ Logistics to work on site and the Grand Chief had this line and it was kind of like, "We issued the challenge to you, the mining company, to hire our people, the northern people. You've now hired us [Th̩chǫ Logistics] to work on your mine site and that means we have to do it, too. And we accept the challenge." So it's an interesting turnabout. So rather than the old pointing the finger, you know that old saying – when you point one finger at someone, there's four pointing back at you. (Interview 401)[16]

So, in contrast to the "separate development" of previous extractive regimes, Indigenous communities are newly implicated and, thus, newly constrained in their efforts to challenge the diamond-mining regime. In this way, the push for Indigenous businesses serves to silence dissent by tying the well-being of Indigenous communities to the success of extractive capitalist production.

However, while the emphasis on Indigenous participation in the FIFO diamond-mining regime emerges from the contradictory imperatives of recognizing distinct Indigenous relationships to land and the capitalist drive for accumulation and exploitation, this does not deny the benefits Indigenous people have gained from working in community-run companies. For some, Indigenous businesses offer an opportunity to work with fellow residents of their own community and to feel more autonomy over their engagement in an industry that has been imposed upon them. Furthermore, working in the mines is a chance to gain local employment at a time when neoliberal retrenchment of services, the high cost of living, and deepening capitalist penetration into northern spaces is making living in the North without engaging in wage labour increasingly difficult. In both interviews and talking circles, research participants noted the real benefits of the development of Indigenous businesses auxiliary to the diamond industry, both at the individual level (through jobs) and at the community level.

At the same time, even those research participants who appreciate the opportunity to work in Indigenous-run businesses were quick to note the tension between the extractive regime and Indigenous modes of life. For example, Angus, a young Métis man who participated in a community talking circle, said:

> I feel like the diamond mines places this thing over their head, this exit out, this easy answer to things. And it's almost like the vampire disease. Once we have it, we just keep chasing mine sites. And once a mine starts getting low, we start searching for the next one. And I think it's counter-intuitive to our philosophy of being here to co-exist on this land, you know, and maintain it for future generations. We end up making a deal with the devil. (Talking Circle 2)

Angus was raising concerns about the attachment that develops between Indigenous communities and the extractive economy as Indigenous businesses are built around the diamond mines. For him, rather than a short-term economic opportunity, the diamond mines represented a larger threat to the reproduction of northern Indigenous modes of life. Arlene Hache,[17] for her part, articulated the concern that Indigenous organizations were taking up Western labour practices, noting the implications this drift toward capitalism had for intergenerational Indigenous social reproduction. Arlene acknowledged the potential benefits of Indigenous businesses, but then added:

> I find it a sort of sad thing because there are real values that will be lost. Or is not acknowledged, or will just change the nature of community over time. So one of the values is, "I will be there for family and the community." And, you know, the Western wage economy kind of way is, don't bring family to work. Don't bring problems to work. You're here to work for me from 9 to 5 or whatever. So I think there's a real sadness about real community values that are so important being lost and not accepted and, soon, not accepted in the community itself.
>
> So sometimes it's very tough when Indigenous organizations, business organizations, follow the same model as Western organizations. And now, all of a sudden, community expectations and family expectations also are not welcome in Aboriginal businesses and models. So it's something to think about, anyways. (Interview 110)

Arlene's and Angus's concerns point to the threat that Indigenous labour may be reorganized materially and ontologically as a result of the imperatives of the diamond-mining regime. This is not to say that Indigenous peoples cannot, should not, or do not participate in or benefit from the diamond-mining regime. Rather, it is to acknowledge a real tension between land-based social relations and the capitalist imperative for accumulation, intensified by the FIFO extractive structure. It is when we examine capitalist production in the FIFO diamond-mining regime as one department of production in relation to subsistence and social reproduction in the mixed economy that the violent effects of this reorganization manifest themselves. The next three chapters take up this task, examining northern Indigenous women's multiple labours as they relate to the diamond-mining regime.

Conclusion

The diamond-mining regime emerged in the context of a rupture in its predecessor, the gold-mining regime, and was structured through the neoliberal imperatives to emphasize variable over fixed capital as well as through the legislative, policy, and ideological gains made in the course of Indigenous organizing in the twentieth century. The result was new attention to Indigenous communities and people in the NWT. While this has included consultation with communities and some revenue distribution to Indigenous organizations, the diamond-mining regime's primary approach to Indigenous peoples is as potential workers and business operators. This emphasis is informed by an approach to Indigenous peoples detached from the material distinctness of their demands: demands for autonomy over and access to land, demands for the ability to pursue their modes of life. For the diamond-mining regime, "being Indigenous" is largely an identity-marker amenable to incorporation in mining operations, rooted in liberal capitalist individualistic rights-based structures (i.e., Indigenous people have the right to profit from the diamond mines, as long as they participate in ways dictated by the Canadian state and the needs of private capital). The third part of this book, which draws primarily on interviews, talking circles, and focus groups, takes an expansive approach to discussing northern Indigenous women's experiences with the diamond industry. Using the expanded conception of production established in chapter 2, then, the focus of the following chapter is on capitalist production (work at the mining camp itself), whereas that of chapter 7 is on day-to-day social reproduction, and that of chapter 8 is on subsistence and the intergenerational social reproduction of northern Indigenous modes of life.

PART THREE

Indigenous Women's Labour and the Diamond Mines

6 Time, Place, and the Diamond-Mining Regime

I mean, the biggest reason I wanted to talk to you is that, for me, the diamond mines coming in, it was like that cloud cover coming over [points to big dark clouds on Great Slave Lake], only not so drastic. We were like frogs in the water and it was heating up.

– Interview 105

The diamond mines exemplify not only the new political imperatives to engage Indigenous peoples in northern extraction but also the economic imperatives of neoliberalism for leaner operations (Peck 2013), demonstrable in the FIFO extractive model. Thus, in examining wage labour under the diamond-mining regime and the experiences of women, especially Indigenous women, working for the diamond mines, I focus on the articulation of Indigenous engagement in the context of FIFO operations. The FIFO model, newly popular in global extraction, attempts to evade the fixity of resource extraction (i.e., that extraction must occur where the resources are) and its consequent need for investments in fixed capital by temporarily importing workers from their homes to the mine site for the duration of their work. In the context of the diamond mines, this is usually fourteen days of twelve-hour shifts. The FIFO model, which I discuss in the first section of this chapter, is thus a spatial articulation of capitalism's gender division between masculinized capitalist production and feminized social reproduction. While the day-to-day social reproduction of mine workers is performed at the mine site through a gendered and racialized employment hierarchy, the intergenerational social reproduction of Indigenous communities is feminized and externalized.

Thus, diamond mining has brought a new gender dynamic to wage labour in NWT Indigenous communities. At the same time, one of the implications of the FIFO model is that it reduces the need for labour power, as a result of both the intensity of shifts and the flexible, just-in-time nature of employment that has accompanied the turn toward FIFO. This means that, while diamond mining weighs heavily on the economic development ideology shaping NWT governance (particularly at the federal and territorial levels), in Section II, I argue that this is not the result of strong employment linkages. Instead, it relates more to the affective need for extraction, reproduced by the ideological emphasis on masculinized (extractive) labour over feminized – social, educative, and care – labour. The latter labour sectors provide far more jobs than mining in the NWT labour market. At the same time, the need for extractive labour is by no means *only* affective. The diamond industry does offer relatively high-wage employment in the mines themselves and in auxiliary industries, and the FIFO model makes this form of employment newly accessible to those do not wish to permanently move from their community. The impact of diamond-mining jobs plays out through high levels of economic inequality in the NWT mixed economy and the ups and downs of the extractive market.

Thus, Indigenous women's labour at the mine site is shaped by the spatial and ideological separation between (masculinized) capitalist production and (feminized) social reproduction, as well as by the new emphasis on recruiting northern Indigenous people as a potential workforce. In Section III, I argue that Indigenous women's experiences working in the diamond mines, and the violence some Indigenous women face at work or as a result of their work, are structured through a gendered settler colonial employment hierarchy. This hierarchy plays out through the contradictory imperatives for the diamond-mining regime to target northern Indigenous women as a potential local workforce and at the same time feminize and externalize social reproduction, and racialize and marginalize subsistence, thereby limiting Indigenous women's capacity to engage in diamond-mining labour in a steady, long-term fashion. These contradictions are largely resolved for the FIFO diamond-mining regime, which treats some Indigenous women as what Melissa Wright (1999) calls '"untrainable" workers, a feminized and racialized category that allows for intensified exploitation of workers who are imagined as short-term employees.

I suggest that the violence faced by Indigenous women who attempt to work in the mine (and the relative failure of industry and Canadian state recruitment and training strategies geared to Indigenous women) speaks to the tension between the Indigenous labourer/entrepreneur imagined and encouraged by the Canadian state and private capital, and Indigenous practices of social reproduction and subsistence in the mixed economy. In other words, the experiences of sexual harassment and gender discrimination described by Indigenous women who worked at the mines are embodied and interpersonal manifestations of the structural violence embedded in the shifts in the social relations of the regional mixed economy generated by the FIFO diamond-mining regime.

I. FIFO in the Mixed Economy

a) The FIFO Turn

The twentieth century saw settler towns growing alongside northern mining developments, but by the early 1990s, mining settlements were no longer the order of the day. Workers expected more infrastructure than the mining generation of early-settler Yellowknife, and building towns had become more expensive, an investment in fixed capital at odds with neoliberal norms related to lean operations. Thus, in Canada and elsewhere (notably Australia), a new extractive model began replacing the mining town – the FIFO mine. FIFO – or, in some cases, DIDO (drive-in, drive-out) – is not new: as a production model mediating the space between work and home, it has popped up in a wide range of industries, from the army to medical services to education. In resource extraction, it first emerged in the offshore oil industry after the Second World War. Markey, Storey, and Heisler (2011) write that "as offshore activity extended further from the beach, daily commuting was no longer feasible and as there were no 'local' communities from which to draw workers or 'local' accommodations available, fly-in/fly-out was the only practical solution" (214). Canada's first FIFO mine was Asbestos Hill in Quebec in the late 1960s. The model gained momentum over the following decades, particularly in the North, where the geography is expansive and construction costs are high. Indeed, between 1982 and 1992, there were no new northern Canadian mining towns built, compared to sixteen new FIFO operations (Markey 2010). Storey (2010) writes that, due to the attractiveness of FIFO's lean and flexible mode of production for remote area extraction, "over the past twenty-five years, and in Canada and Australia in particular, the 'no town' model has replaced that of the 'new town'" (1161).

The "no town" model is a neoliberal move toward investment in variable capital over fixed capital, a move that facilitates the intensified flexibility of the diamond mines. In Marxian political economy, variable capital refers to the amount of value added to a commodity by workers' labour, while fixed capital refers to the physical structures (e.g., a road or a factory) that are used for capitalist production and constant capital refers to the wear and tear on fixed capital required in a given production cycle (Fine and Filho 2010: 34–5). According to Marx, both variable capital and fixed capital contribute to the exchange value of a commodity, but it is only the exploitation of workers' surplus labour (variable capital) that creates new value. Thus, there is a capitalist imperative to replace fixed capital with variable capital where possible.[1] In the neoliberal era, where state and industry alike strive for lean operations oriented toward the immediate needs of capital, an emphasis on variable capital over fixed capital facilitates the increased temporariness of production. Indeed, as discussed in this book's introductory chapter, the contemporary model of capitalism normalizes temporariness (Vosko et al., eds. 2014) and mobility (Walsh 2012), wherein workers are increasingly vulnerable to the fluid transnational flows of capital as well as to the consequent demands for movable workspaces and people (e.g., production moved to places with lower costs of labour, offices moved online to avoid infrastructure costs).

Resource extraction holds a contradictory location in the flexible temporariness that characterizes late capitalism: on the one hand, resource extraction is defined by its temporariness, a temporariness that, as we have seen, has shaped the violent expendability of extractive workers across space and time. However, while tech companies, for example, can to a large degree avoid the constraints of space by contracting workers from so many coffee shops and basement apartments, just as extraction is defined by its temporariness, so, too, is it defined by the materiality of the space it inhabits: the mine must follow the minerals, not the reverse.[2] In avoiding the fixed costs of building a company town by relying on the comparatively low costs of flying, FIFO companies effectively annex space with time. The result is not, as Markey (2010) reminds us, "place-less" development. The mining camp to which people fly is very much a place and, less recognized, so too are the communities, towns, and cities surrounding mine sites to which people fly home. These homes are not in traditional mining towns and thus do not benefit from the same sympathy and support often afforded to mining towns over the course of the ups and downs of extractive practices (McDonald et al. 2012),[3] as the impact of FIFO's short-term contracts and abrupt layoffs are lost in the airspace so regularly traversed. The eroded relationship between extractive company and town, common to FIFO, is exacerbated by the difficulty of union organizing in mining camps (Ellem 2013), but in the context of the NWT, this is mitigated by the strong norms for Indigenous community consultation.

FIFO, then, intensifies the general flexibility and temporariness of extractive labour. At the same time, in its emphasis on variable capital, it is a specific response to the costs of labour and social reproduction in the countries where it is most popular: Canada and Australia. Indeed, it is no accident that FIFO is practised primarily in these two settler colonial states: both are resource-rich, expansive polities where construction costs are high and extractive labourers (traditionally and, to this day, often male, often white) expect relatively high salaries and amenities. In many countries in the Global South, the political economy of neocolonial extraction (the low wages and low health, safety, and living standards attached to extractive projects made possible by unequal international trade agreements [Tienhaara 2011, Gordon and Webber 2008]) makes the spatial fix of FIFO obsolete. In contrast, the NWT, with its low population/land ratio and the challenges the cold, rocky landscape poses to Western forms of infrastructure, intensifies the economic imperatives for FIFO found across Canada. In the NWT, the first large-scale FIFO mine was Polaris, a zinc mine in Nunavut that operated from 1981 to 2002. In an interview, James said that the choice to make Polaris a FIFO mine was a reaction to the massive expenditures throughout the mid-twentieth century in building mining towns and handling the town infrastructure once the mine closed:

I say that Nanasivik was the last mining town to be built, up in Nunavut.[4] If you look at Nanasivik when it was built, the federal government had a 20 per cent stake in it when it was built. And what they did was … community services. So they put in the airstrip and they paid for the dock. They paid for the roads and they paid for the community roads and the RCMP. They paid for a town centre that had things like a post office, a store, and a nursing station. They had a fire hall. And school. And I can remember, I was there when it was built. And they paid for the water sewer. There was a whole utility system and for maintenance. So that was their contribution. So there you've got a case of government getting a financial burden for having a mining town. Because they also have to clean up their stuff … A mining cost. So with fly-in/fly-out, there's an elegance to that. There's a cost saving. There's a "We don't have to go through that," you know? (Interview 401)

All of the NWT diamond mines were developed at a significant distance from Yellowknife and the other towns of the territory.[5] The mines are remote, even by northern standards. Extending existing infrastructure and transporting all the material resources required to build a town would have been far more of an investment than anyone (government or private capital) was willing to bestow on this project. Furthermore, the cost of flying has significantly decreased, making the constant flights taking off from the diamond mines and bringing workers to their homes – which range from communities around the North to locales as far as the Maritimes – a reasonable expenditure for the mining companies. As James noted, the FIFO structure allows workers to take a job at a mine without having to relocate themselves or their families. This, he argued, is particularly important for recruiting northern Indigenous people who do not want to leave their home community (Interview 401).

However, while home communities remain populated (at least half the time) and in place, their gendered social relations have undergone a significant shift under the FIFO diamond-mining regime. Namely, FIFO introduces a spatiality to the assumed separation under capitalism between productive (capitalist) labour and social reproduction. That is, while capitalist production is separated from social reproduction in Western liberal ideology (the public/private realm; the work/home divide; the masculine/feminine) and materiality, FIFO spatializes this separation, placing hundreds of kilometres of Subarctic land between the site of capitalist production and the site of social reproduction while obscuring land-based relations that operate antithetically to the Western work/home divide. This separation has multiple implications for Indigenous women's labour, both on and off the mine site. Because FIFO diamond-mining labour usually involves workers living at camp for two weeks at a time and working twelve hours a day for fourteen days straight, the daily social reproduction of workers is made possible by paid labour at camp (cooking and housekeeping, for example, most often performed by Indigenous women), labour that is largely obscured and devalued. While (predominantly male) workers are at camp, daily social reproduction at "home" – especially child care – is feminized and naturalized, and subsistence is largely invisibilized. The feminization of social reproduction has been facilitated, in part, by twentieth-century state interventions into Indigenous women's labour and Indigenous social reproduction (discussed in chapter 3). As the subsequent chapters will show, while many female spouses of mine workers take on increased household loads, the work of intergenerational social reproduction of Indigenous communities is hampered by the new long-term separation of mine workers from their home communities, and the workers' responsibilities to their community and kin are, out of necessity, taken up by other family members or friends (often women). Thus, due to their intensified community and household responsibilities, many women are unable to access or (if accessed) sustain employment at the diamond mines. Gendered lack of access to diamond-mine employment is particularly salient given the status that mine employment has in the regional political economy.

b) FIFO and Work in the NWT

As with many industries in more rural or remote jurisdictions, in debates around northern extraction, jobs are often the carrot used to gain favour for an industry's existence. The disciplinary power of the promise of new jobs and, conversely, the fear of unemployment is, arguably, the primary tactic used contemporaneously by the capitalist class (both political and economic) to justify new development projects – and in the North, extractive projects in particular. As the GNWT (2014a), writes, "the GNWT recognizes that resource development is the foundation of the NWT economy." Certainly, this emphasis on wage – and, particularly, extractive – labour denies subsistence production its role in having sustained northern Indigenous communities over the centuries. However, even within the sphere of capitalist employment, the powerful rhetoric around job creation through mining obscures the diversity of employment in the NWT. Yellowknife is adorned with flags identifying the town as the "Diamond Capital" (see Image 1), yet surprisingly few people in the Yellowknife region, and the NWT at large, actually work for the diamond mines. Indeed, the diamond mines have minimized their workforce needs through long shifts, long days (Gibson 2008), and short-term contracts. This, combined with the industry's ability to hire from outside the territory, leads to weak local employment linkages, especially compared to the gold mines. Figures 2 and 3 draw on GNWT data to compare the mining sector's share in the territory's GDP distribution and labour market, respectively.

As Figures 2 and 3 illustrate, while extractive industries (the vast majority of which are diamond mines) dominate territorial GDP distribution, they employ a relatively small proportion of the NWT's population. Public administration employment, in particular, overshadows extractive employment, an important corrective to masculine narratives espousing the primacy of mine employment (see below). Figures 2 and 3 represent territorial statistics, as these data were not available at a municipal level. However, it would be safe to assume that in the Yellowknife region – where most non-mining-related wage labour is performed – the proportion of people working in mining compared to other forms of wage labour would be even less.

The mines measure employment in person-hours, counting the number of hours worked rather than the number of people employed. This

Figure 2. Percentage (%) of territorial GDP by industry (2019) (2021 NWT Bureau of Statistics).

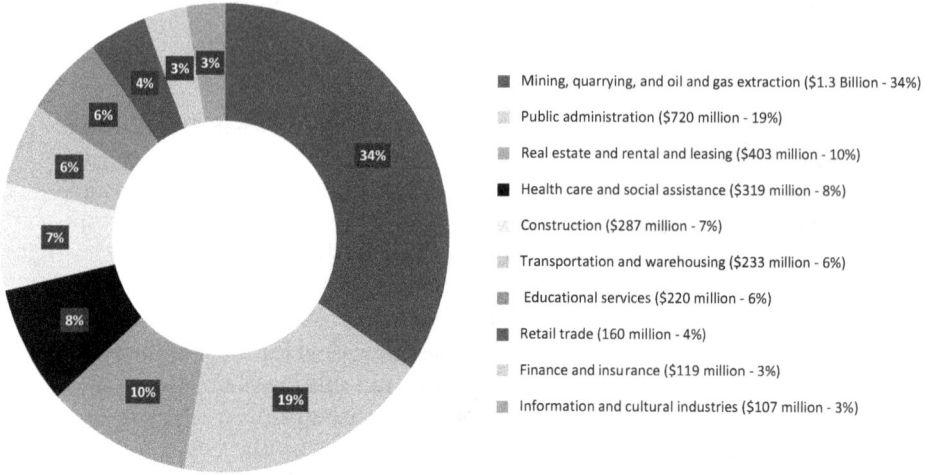

- Mining, quarrying, and oil and gas extraction ($1.3 Billion - 34%)
- Public administration ($720 million - 19%)
- Real estate and rental and leasing ($403 million - 10%)
- Health care and social assistance ($319 million - 8%)
- Construction ($287 million - 7%)
- Transportation and warehousing ($233 million - 6%)
- Educational services ($220 million - 6%)
- Retail trade (160 million - 4%)
- Finance and insurance ($119 million - 3%)
- Information and cultural industries ($107 million - 3%)

Figure 3. Employment, by class of worker. Northwest Territories, 2001–2019 (2021 NWT Bureau of Statistics).

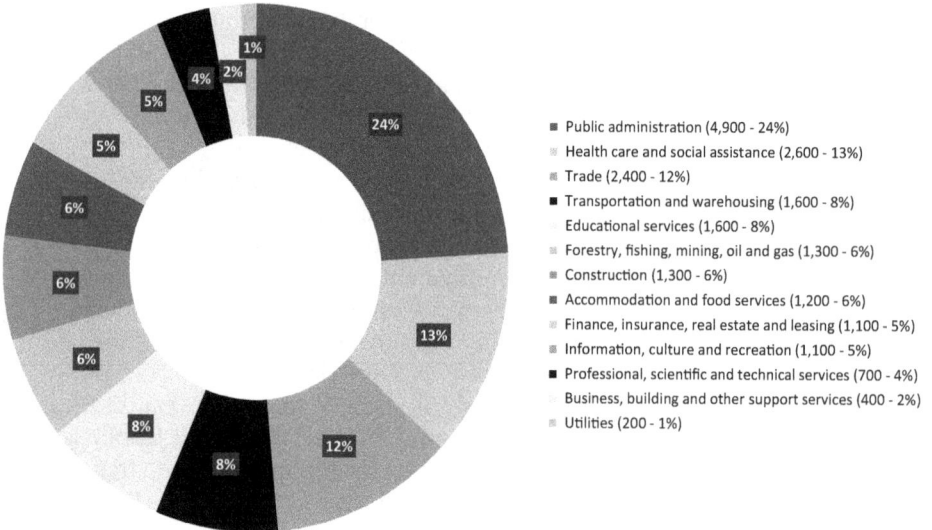

- Public administration (4,900 - 24%)
- Health care and social assistance (2,600 - 13%)
- Trade (2,400 - 12%)
- Transportation and warehousing (1,600 - 8%)
  Educational services (1,600 - 8%)
- Forestry, fishing, mining, oil and gas (1,300 - 6%)
- Construction (1,300 - 6%)
- Accommodation and food services (1,200 - 6%)
- Finance, insurance, real estate and leasing (1,100 - 5%)
- Information, culture and recreation (1,100 - 5%)
- Professional, scientific and technical services (700 - 4%)
  Business, building and other support services (400 - 2%)
- Utilities (200 - 1%)

mode of data collection, which tracks labour-power rather than people, is itself an articulation of an emphasis on variable capital. The person who is expending labour is lost in this form of data collection, which thereby obscures the nature of employment (i.e., Who is working what hours? And what is the duration of their employment?). According to the GNWT (2019a), the cumulative (from 1996 to 2017) number of person-years in northern employment supplied to northern people across the four diamond mines was 27,612, approximately 50 per cent of total employment. Of these northern person-years, 13,741 went to northern Indigenous employees, or 24 per cent of total employment, thereby meeting territorial targets. In 2017, for example, operating mines provided 1,592 direct northern jobs, of which 819 went to northern Indigenous people. The reports from which these data were gathered do not disaggregate by sex, nor do they identify non-Indigenous minorities. However, according to the NWT and Nunavut Chamber of Mines (2017), female employment runs at around 14 per cent of total employment. In its socio-economic reporting on mining labour force activity (2019a), the GNWT also accounts for jobs in industries auxiliary to the industry: some are clear-cut examples, such as construction and transportation companies servicing the mines. The impact of the mines on sectors like retail and tourism is speculative.

The diamond mines provide employment that is generally characterized by research participants as high-wage or, at least, "higher than I would have gotten somewhere else." For example, Iris had done housekeeping work in town before being hired as a housekeeper up at the mines. She said she started working at the mines because the pay was better (Interview 101). However, while there is an assumption that mines bring "good jobs," for those who do work in a mine, there is significant variety in job type, form of employment (contract or full-time, with varying levels of security), and wages. The discrepancy in job types is exacerbated by the limited union presence in the diamond mines, a major difference between the gold mines and the diamond mines. Even among those diamond mine workers who are unionized, the union does not play the activist role it did during the gold-mining days. When asked about the difference between union organizing during the gold mine era and union organizing at the diamond mines, Mary Lou Cherwaty, the President of the Northern Territories Federation of Labour, replied that[6]

> from a union perspective, there's a huge difference. Because you don't have access to the worksite. If you go out to Giant or Con [gold mines] or something where your family and community live close to the worksite.

And people drive or walk to the workplace. As opposed to now, where people get on an airplane to go. So the only contact you can have with those workers is to go to Yellowknife to the airport or wherever they're flying from. Some of those workers, even though they're your members, you don't even get contact with them. Unless you go in on a visit through the employer's permission. On their flight to the mine.

R: And it [the flight] would have to be through the employers?

M: Yes, because you need to be on their flight manifest through the airline. But even then you have construction contract workers and you've got your housekeeping services, which is a separate employer. They're all broken out. So you get an airplane of people going to work on a shift and they all could work for eight different employers. (Interview 204)

Cherwaty's analysis of the challenges of union organizing is a spatial one: when the site of capitalist production in the form of diamond mining is separate from the site of social reproduction and access is limited to those invited by the company themselves (i.e., who are on the private flight manifesto), the site of capitalist production arguably falls outside of the influence of social organizations, or at least is more difficult for them to influence. Indeed, echoing concerns coming from Australian FIFO literature (Ellem 2013), Cherwaty speaks to the near-impossibility of union organizing at a FIFO site, noting the lack of knowledge of and engagement with the unions on the part of the membership.

This sentiment was illustrated by a few research participants (Indigenous women who had worked in the mines), who did not know whether their position at the mine was unionized or not, sometimes due to uncertainty over whether they were employed by the mine itself or by a subcontractor, an uncertainty created by short-term contracts and lack of union access. Two community workers explained that it is difficult to help workers access their rights because it is often not clear who employs them. According to Jane, "We also have the big problem with the subcontracting. The issues around work … It's hard to know who they work for, who's HR" (Interview 201). This lack of clarity in employment relationships makes it difficult for workers to lodge complaints or to access their rights. Clarice, a Dene woman working at a mine, experienced severe workplace harassment. She explained to me that she did not know that she was a union member and that she could access union resources. In the gold mines, the jobs had been housed under one employer, whereas under the present diamond-mining regime, a great deal of work is farmed out to sub-contractors. Because sub-contractors are not party to SEAs or IBAs, they are not under the same labour obligations as the diamond mines. A few interview participants

Figure 4. Rates of employment (GNWT 2018a)

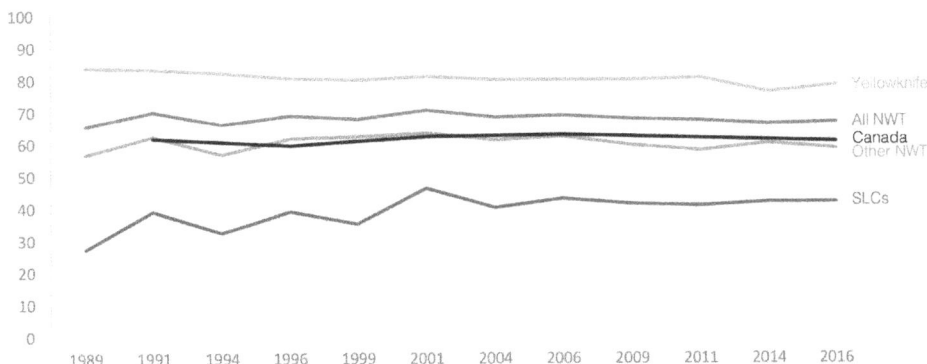

Sources: 1989, 1994 and 1999 NWT Labour Force Survey; 2004, 2009 and 2014 NWT Community Survey; 2006 Statistics Canada Census; 2011 Statistics Canada National Household Survey.

described working directly for the diamond mine as the "good job" compared to working for subcontractors, a characterization that was echoed in the community focus group.[7]

II. Affective Need and Inequality

How, then, has wage labour changed as a result of the diamond mines? Figure 4 tracks employment rates through the development of the diamond mines.[8] Recalling that the first diamond mine opened in 1998 and the second in 2003, it is notable that the level of employment remained relatively stable throughout this time period. Indeed, this graph illustrates the long-standing high level of employment in the NWT. Given that those who are characterized as "unemployed" are often Indigenous people engaged in the subsistence economy, it is arguable that employment concerns in the NWT are, for the most part, not related to a lack of jobs, but are more likely to be rooted in concerns related to the relationship between wage labour and Indigenous modes of life, job quality, and the ability of wages to meet the increasingly high cost of living in the NWT. As evident in the graph, the one exception to the steady employment rates is employment in small local communities (SLCs), which has seen a significant increase as a result of the diamond mines. This is because the FIFO structure of the mines allows people to remain in small communities. SLCs have much higher rates of participation in subsistence activities than larger communities; this is labour that is not tracked in these GNWT data.

However, while employment numbers – in the territory, overall, and in the region of study – have not risen significantly since the opening of the diamond mines (noting, certainly, that the opening of the diamond mines coincided with the closing of the gold mines and, thus, the spike in diamond employment would be lost in aggregate extractive and territorial data), and while significantly more NWT residents work in non-mining jobs than are employed in the mining sector, it does not follow that the diamond mines have inserted themselves quietly into the NWT capitalist labour market. Instead, the diamond mines have perpetuated and intensified an affective "need" for extraction that has shaped the labour market in the NWT mixed economy (through, for example, training and government resource allocation), while deepening economic inequality. I discuss these two ramifications in turn.

State (both federal and territorial) and business interest in reproducing an extractive economy is clear, but the social reproduction of extractive "need" is multi-scalar: it is enacted through the free-entry regime perpetuated by elite politics, to be sure, but also by decisions at all levels of governance regarding where education and training resources should go and what businesses and initiatives should be supported, as well as in the discursive culture that makes a place a "mining town." Our interviewees and focus groups often spoke of the importance of training more young people for the diamond industry, and indeed, huge amounts of government and community organizational resources and time have been invested in this training. For example, since its inception in 2004, the NWT's Mine Training Society has received $50.8 million in funding and in-kind support from federal and territorial governments, and from Indigenous and industry partners (NWT and Nunavut Chamber of Mines 2017), with $31 million coming from Employment and Social Development Canada. By 2017, the diamond industry had passed its peak, yet the federal government still awarded the Mine Training Society an additional $7.4 million to train Indigenous mine workers (Trochu 2017). Investments in the Mine Training Society have been complemented by community-specific training initiatives delineated in IBAs, apprenticeship programs targeting Indigenous peoples managed by the diamond mines (in collaboration with the Mining Training Society), literacy and equivalency programs aimed at helping community members meet literacy and education requirements for mine employment, and, notably, the Northern Women in Mining, Oil, and Gas Project, a $1.3 million project aimed at training and recruiting Indigenous women for the extractives sector (Carey Consulting and Evaluation 2011).

These programs, of course, do not emerge from nothing, and they sometimes tap into resources that had been allocated elsewhere. Lois Little (2007), a northern community researcher and advocate, has noted that the high value placed on masculinized mine labour diminishes the resources for feminized social, educative, and health labour. Perhaps the most explicit example of the reallocation toward extraction was the attempted cancellation in 2017 of the social work program at Aurora College, the NWT's only college. The NWT's education minister at the time, Alfred Moses, announced that that program had been cancelled due to cutbacks and low enrolment. Fortunately, as a result of community pressure, the cancellation was revoked, but the vulnerability of the small social work program to cutbacks sits in contrast with the new $10 Million Aurora College Mine Training Centre opened in 2019.[9] Social workers are in high demand in the NWT, and these positions are often taken up by new graduates from southern Canada. A review of the social work program commissioned by the GNWT emphasized its importance for training local, Indigenous social workers. That report, which was buried until progressive MLA Julie Green filed an access-to-information request (Beers 2018), recommended changes to the curriculum to make the social work program culturally relevant, but certainly not the program's closure (Edwards 2019). Given that far more NWT residents, and especially far more women, are employed in various forms of government and non-government service and care work than in the diamond mines, the steering of resources to mine training rather than the social work program reflects the government's prioritization of reproducing an extractive state over the demands of the regional labour market and, more perniciously, the needs of the community.

Thus, the diamond mines have extended the expectation for extractive employment initially established by the gold mines, even while introducing a new precariousness to mine employment. The emphasis on variable capital under the FIFO diamond-mining regime has not translated into more jobs, but rather into an *affective concern* for extractive jobs, one that privileges masculinized over feminized forms of work. Rebecca Scott's (2007) analysis of Blair, a coal-mining town in West Virginia, identifies a similar socialized reliance on extractive employment, and her analysis of the gendered implications is instructive. Like Yellowknife, Blair sees itself as a mining town: "Despite the fact that, as of 1990, retail made up almost as large a percentage of total employment in Boone, Logan, and McDowell counties as coal mining and WalMart is the fourth largest employer in Logan County, this type of low-wage employment was not discussed in my interviews. When people talked about work, they were talking about men, particularly family men-breadwinners" (Scott 2007: 489).

Scott's observation speaks to the way in which (predominantly male and entirely masculinized) extractive employment exacerbates a male-breadwinner ideology and the consequent invisibilization of women's paid and unpaid labour. Just as in Blair, many of the female partners of male mine workers in the Yellowknife region hold jobs of their own (in fact, *all* of the seventeen interview participants who were initially identified because their partners worked in the mine were themselves employed in wage labour, a trend that was also encountered in the two community focus groups). Unlike in Blair, many women working in non-mining jobs in Yellowknife hold high-paying government jobs, a dynamic obscured by the overrepresentation of mining jobs in the region's self-perception. The overrepresentation of mining jobs also obscures the burgeoning service sector in the area around Yellowknife and, through NWT participation in the temporary foreign worker program (GNWT 2015b), its increasingly racialized character. The emphasis on work connected to diamond mining at the expense of the many forms of wage labour within the mixed economy speaks to the concerns outlined by Angus about the "vampire curse" of the mines, or the perceived need for the diamond industry. Indeed, many participants noted that the expectation that men "provide" for their families has intensified in Indigenous homes and is increasingly synonymous with engaging in wage labour, rather than contributing more interdependently through subsistence or social reproduction. The flipside of the expectation that men "provide" through wage labour is that women, regardless of their employment status, are made responsible for household labours, a dynamic we explore in the next chapter.

This distorted emphasis on extractive employment obscures the multiple labours that make up the mixed economy and blocks the imagination of what *could* be. Indeed, while the NWT, like the North in general, is host to Indigenous groups and organizations that combine subsistence and wage labour (Southcott 2015) alongside a much higher proportion of socially oriented forms of wage labour than in the rest of Canada, the affective emphasis on extractive employment protects the extraction-as-development status quo (at both territorial and federal levels). Community organizations, responding to government funding parameters, have shaped their training and education resources to meet the needs of the diamond-mining regime, and with diamond mine closure looming and the GNWT providing no real alternatives, even those research participants most critical of the diamond mines expressed concern about diamond mine closure. The violent social restructuring of

Indigenous modes of life throughout the twentieth century (recall the triadic relationship between treaty payments, resettlement, and mandatory government schools) created a new vulnerability to the ups and downs of extractive production in NWT Indigenous communities. And while historically, Canadian state programs did not always succeed in steering local Indigenous people into extractive labour, they did result in painful ruptures in the daily and intergenerational social reproduction performed in Indigenous communities, ruptures that people are seeking to heal from at the same time that they face the new demands of the FIFO diamond-mining regime.

The FIFO regime articulates through this rupture, often exacerbating existing inequalities. Official poverty rates are high in the NWT. However, I approach the concept of poverty with caution. Especially given the territory's cold climate, lack of material resources can be deeply damaging (physically, interpersonally, and emotionally) and must be addressed. At the same time, state (particularly federal) economic indicators often equate a disengagement from capitalist production with poverty, thereby obscuring the ways in which subsistence production can meet the day-to-day needs of people living in and around Yellowknife. Far from painting the NWT as a place of some sort of inherent or ahistoric *lack*, I focus on *inequality* and the gap between financial assets and the currently high cost of living in the North.

The NWT's high GDP – twice the Canadian average – is the result of the diamond mines (and, previously, other extractive regimes) and high-paying government jobs; consequently, household incomes are vastly polarized. In Yellowknife, a full 50 per cent of households have an annual income of $100,000 or more, while 12 per cent have a household income of less than $30,000 and 3 per cent have an income of less than $10,000 (Wilson 2009: 9). The proportion of households living on incomes below $10,000 and $30,000 is less than the Canadian average, but the cost of living is significantly higher in Yellowknife compared to other Canadian cities. Indeed, the GNWT estimates that the cost of living, when compared to that of Edmonton (as a marker for an average southern Canadian city), is 120 to 125 per cent higher in Yellowknife, 125 to 130 per cent higher in Behchokǫ̀, and 140 to 180 per cent higher in smaller NWT communities (GNWT 2013a). Because the GNWT economic data do not disaggregate Dettah and N'dilo from Yellowknife, their distinct socio-economic conditions are not reflected in territorial socio-economic data. In Behchokǫ̀, 6 per cent of households have an income under $10,000 and 30 per cent of households have an income under $30,000, while 31 per cent of households have an income over $100,000 (Wilson 2009: 9). Given that Behchokǫ̀ is a community with both high participation in the diamond mines and a strong engagement with the mixed economy, it is not surprising that income inequality in that town is very high.

Given that the settler high-wage model developed in and around Yellowknife through the gold mines, and was bolstered with the introduction of government jobs in the 1960s, wage inequality is not new to this region. However, in Yellowknife, the diamond-mining resource boom has only made things harder for the segment of the population that does not fall into the "50%," that is, the 50 per cent of Yellowknife residents with annual salaries over $100,000. Northern store-bought food costs, for example, can be prohibitive (a much greater concern in smaller communities, but an issue throughout the territory).[10] And average rent for an apartment in Yellowknife increased by 50 per cent after the diamond mines opened (Falvo 2011: 249). As Julia Christensen (2014) notes:

> Recent economic growth due to resource development has meant an unprecedented private rental-housing crunch; low vacancy combined with the high incomes from government and industry employment motivates an expensive, exclusive private rental market. Meanwhile, efforts to move away from its role in public housing provision has led the GNWT to increase its scrutiny of public housing residents, including the recent implementation of a "no tolerance" policy on arrears that led to widespread public housing evictions in 2012 as well as selling off some of its stock in larger market communities, [like Yellowknife]. (813)

In Yellowknife, where most of the territorial wealth is concentrated, the strain of high-cost rentals is more acute than in smaller NWT towns, where housing is almost exclusively government owned and operated.[11] The rates of street-affected people in Yellowknife demonstrate the intensity of the inequality experienced in this town of 20,000. According to Christensen (2014), visible homelessness first appeared in Yellowknife at the same time that the diamond mines opened (805). Currently, 5 per cent of the municipal population uses a shelter at least once a year. That rate is five times higher than in the average Canadian city (Falvo 2011: 5). Rates of homelessness for women are also significantly higher in the NWT. The 5 per cent rate holds for women as well as men, whereas in most of Canada, rates of homelessness among men are higher than rates of homelessness among women.

The division between the "haves" and "have-nots" does not fall neatly along Indigenous/settler lines: many Dene, Métis, and Inuit residents of the area around Yellowknife hold middle- to high-income jobs in government offices, the diamond industry, or auxiliary industries. However, poverty is certainly racialized, having been shaped by centuries of settler colonial dispossession and, more recently, by the unequal incorporation of migrants into the northern economy. The number of visible minorities in Yellowknife has risen steadily over the past ten years (Little 2010), in large part due to enthusiastic participation by the region in the temporary foreign worker program, which has been used to fill high vacancy rates in low-paying service sector jobs. The high cost of living in and around Yellowknife makes living on these low wages very challenging. Indigenous people make up the majority of those experiencing high levels of poverty and homelessness. For example, between 90 and 95 per cent of shelter users are Indigenous (Christensen 2014: 806). This racialized inequality shapes experiences of the diamond mines, as well as the structural and embodied violence associated with the mines.

III. Indigenous Women and the FIFO Diamond-Mining Regime

Given the history of stable, high-paying employment in the extractive industry for – predominantly – white settler men living in Yellowknife, for some residents of the region, the diamond mines offered a welcome continuation of this form of employment at a time when the region's political-economic future was violently uncertain. The diamond mines also offered a new form of high-wage employment for Indigenous communities previously excluded from the extractive regime. At the same time, the diamond mines are marked by new forms of precarious labour as a result of increased subcontracting, short-term contracts, and non-unionized jobs. Furthermore, because of the FIFO structure, the daily social reproduction of workers now occurs at the mine site. As such, the diamond mines employ workers to perform this labour (largely cooking and cleaning). This is employment of a different quality, pay scale, and gendering from construction, pit work, or trades. Certainly, women – especially Indigenous women – participate in the diamond-mining regime far more than they did in the gold-mining regime, but where and how do Indigenous women fit (and not fit) into the extractive employment structure? And what do their experiences within the diamond-mining regime tell us about the material and ideological contradictions at play in the restructuring of labour to meet the demands of capital? And what are the violent implications of these contradictions?

Like men, women occupy a wide range of roles in the diamond mines, and the experiences shared by research participants (most of whom were Indigenous women) were, of course, divergent, both within and across subjectivities. In interviews and focus groups, some women spoke of the financial stability afforded to them by work in the diamond mines, while others spoke of the deep insecurity. Some spoke of the empowerment they felt as a result of the new employment opportunities offered by the diamond mines. For example, Jillian, a white northerner, said, "I made enough money to buy my own place before I was in a relationship with my husband and be really independent that way. It allowed me to travel all over the place and really enjoy my time like that. However, as I got older, my priorities changed and then you have to make decision[s] about whether you want to stick with it or do something else in town" (Talking Circle Two). Others spoke of their struggles to move out of low-wage, insecure jobs at the mines. For example, Doris, a Dene woman who worked as a housekeeper at the mines, said, "It's not a high-paying job being a housekeeper. I wanted a job like everybody else was doing. I mean, why can't I get hired? You know? And I would have got a secure person to sit with the kids, if I could afford it" (Interview 111).

Given the small and purposive nature of sampling for this project, the different experiences expressed by research participants should not be taken as representative of the experiences of all Indigenous women working in the diamond mines. Rather, I draw upon the specific narratives shared to illuminate the competing political and ideological imperatives that have shaped the diamond-mining regime. To some extent, Indigenous women are recruited for employment through the same imperatives as their male counterparts, as constituents of the new approach to Indigenous participation, outlined in the previous chapter. At the same time, the diamond-mining regime relies upon a racialized marginalization of subsistence, and an externalization and feminization of social reproduction, with the exception of the day-to-day social reproduction performed on site. As a result, recruitment of Indigenous women for long-term employment has been deeply coloured by the newly individualized responsibilities for household and community social reproduction these women face. These processes, however, have not resulted in non-participation, full stop. Rather, the diamond mines rely on a handful of Indigenous women as "ideal" employees, representative of the diamond-mining regime's responsible approach to extraction, and on the labour of many more Indigenous women – however temporary – to fill the least desirable and lowest-paid jobs at the mine.

In interviews with both government and business industry officials, a common theme emerged in discussions of gender and the mines: northern women, and in particular, northern Indigenous women, were referred to – not ironically – as the untapped resource of the North, or, as James put it, "a big opportunity" (Interview 401). As discussed earlier, a local workforce has many potential benefits for the mining companies: more loyalty, lower transportation costs, and positive public perception, as well as the potential avoidance of protest or legal trouble. However, the existing diamond mines have largely exhausted the segment of the northern population that both wants to work for the mines and fits its model of "employability." Many northerners do not want to work for the diamond mines, and others are barred from employment by education requirements. As Emma put it, "it's time to have an honest conversation. Because there's normally a reason people aren't working in the mines. It's normally a criminal record, addictions, or a life skills issue. Or they don't want to be there" (Interview 402).[12] Recruiting Indigenous women is a way to access a new pool of local people who aren't currently employed by the diamond mines. Furthermore, for an industry that relies, to a certain extent, on a perception of "responsible extraction," recruiting Indigenous women to work in the trades is a quick way to achieve this. This strategy is in line with global corporate social responsibility trends, wherein women are the ideal target employment group, for they are seen as more dependable workers and the very act of hiring them, especially in traditionally male-dominated industries like resource extraction, is deemed responsible.[13] In their reporting, both government and the mining industry (GNWT 2019a; NWT and Nunavut Chamber of Mines 2018) have showcased Indigenous women in leadership and technical roles at the mines. Indeed, these women should be lauded for their achievements and expertise; however, their individual achievements should not be taken as evidence of the gendered employment norms in the mines.

Recruiting Indigenous women to work in the diamond mines is not new. Strategies for recruiting, training, and hiring women have been in discussion since the early days of the diamond mines. The Northern Women in Oil and Gas Project, for example, which operated between 2007 and 2010, was funded by the diamond mines and administered by the Status of Women Council of the NWT. While it provided free comprehensive training, the project found limited success in training women who would go on to retain jobs in the diamond mines (Carey Consulting and Evaluation 2011). Prior to the project, community workers identified the barriers participants in it perceived to entering the mining workforce. Before beginning their training or jobs at the mine, the participants identified gendered stereotypes about mining labour (e.g., mining is a "man's job") as the biggest barrier. After the project was completed, in discussions with participants about the training program and the employment that followed, the experience of discrimination and sexual harassment proved to be a common experience. In a joint interview, two community workers had the following discussion:

> JANE: So that was a big factor for, even if they were single, why they didn't want to go there. So not only were they in a men's world, what they perceived as a man's culture, but it was not safe for them.
>
> HILARY: And going home, as well. A lot of women felt that they were stigmatized by that. "Why would you want to go work with all men? What kind of girl are you, anyway?" So I think they got a lot of that, even from their own families, thinking it wasn't a proper's place [...]
>
> JANE: There's also issues, you know, all different issues, clothing not fitting. Bathrooms not fitting, you know. Just the culture of, I don't know how to explain it.
>
> HILARY: Some of the women felt they just weren't wanted there. And that, you know, the old-school trades people, the people working there, were just going to make it as miserable as they could for the women working there. They knew they had to accept them, but they knew they also had the power to make it as unpleasant as possible so they might just leave. (Interview 201)

While gender discrimination was an expected barrier at the outset of the project, women's responsibilities for community and household social reproduction were not identified by participants as a potential barrier for women working in the mines. However, upon completion of the training and after working at the mine (or attempting to work at the mine), program participants reported that the primary barrier to diamond mine employment was that it was incompatible with their responsibilities in their communities and homes (Carey Consulting and Evaluation 2011).[14] Indeed, the FIFO structure requires workers to be highly unattached, or at least periodically detachable from their home responsibilities. While the worker's immediate needs are met at the mine site, broader and more long-term social reproduction, including the work to sustain kin and community social relations, is externalized.

The findings of the post–Northern Women in Oil and Gas Project assessment were largely echoed by research participants in focus groups and interviews. While particular experiences working at the diamond mines diverged, all of the women interviewed identified their gender identity as shaping their work experience significantly. For example, Asha, a northern woman of colour, said that "if you're a woman, it's very apparent you're a woman" (Interview 118), and Harriet, a white woman who had moved up from southern Canada for her position, said, "I just think as a woman you can't BS. There's less room because people are scrutinizing you a lot more" (Interview 117). Women's experiences of their gender location at the camp ranged from discrimination that they felt they could "handle" to sexual harassment and discrimination that led to job loss, ill health, and deep emotional and interpersonal distress. Like Harriet, Jillian, a white northerner who worked as a tradesperson in the mine, largely spoke positively about her experience there. At the same time, she noted that there were no women's bathrooms in the underground pits, offering this as an example of the limited structural accommodations for women. In this vein, a few women in one talking circle discussed the problem of speaking to their male supervisors about medical problems or going to the all-male nurse practitioners for medical attention. The few women who articulated their gendered treatment as an issue, but not a particularly difficult one, all recognized themselves as aberrations. For example, Jillian said:

> Thankfully I worked with a really respectful group of men. There were some that would make the odd comment once in a while, but I never felt threatened by them. But I'm really kind of bossy and loud, so I wouldn't put up with a lot. But even with my personality, it's really intimidating. In some situations, you don't want to speak up. Like I said before, you want to be accepted and respected by your male co-workers, so there are times when you sort of choose your battles and sometimes that's a good thing, but sometimes you don't say things you want to say. (Talking Circle Two: 2014)

For many women, the result of discrimination and harassment at work, and/or the incompatibility of diamond-mining work with their community and household responsibilities, was a history of on-again/off-again temporary employment at the mines. It is noteworthy that many of the Indigenous women participants – who were the majority of both interview and focus group participants – who originally identified themselves as partners of mine workers had also worked in the mines themselves. Most of them had been employed for short, sporadic periods of time. Usually they had been employed under short-term contracts; that, or they had been fired or laid off, or they had left because they felt unsafe, unwelcome, or unhappy at the mine site or because of their community and household responsibilities. For example, Iris had worked at a few of the northern diamond mines on and off for more than a decade. Her work history at the mines was consistently interrupted by responsibilities in the home or community, or by illness. She had left her first job at a diamond mine after six months to look after her sister and her sister's child in a time of need. Iris's commitments to kin and community at the expense of her engagement with capitalist production are an example of the ways in which the women interviewed prioritized their community relations over the demands of the diamond mines. When we spoke, she was planning to reapply to the mines, and she joked:

> This time, I'm gonna stick with it. I tell all my relatives and my friends, you guys can't die, I'm gonna go to work. And they say, "Okay, we won't die." And I say, "Let's see how long you can hold out." I say, "Something can happen, but not while I'm working." But I know that won't work. (Interview 101)

While short-term, insecure labour is characterized as an aberration from the extractive employment model, with diamond mines priding themselves on offering "good jobs," I argue that the FIFO diamond-mining regime operates through a gendered and racialized hierarchy of ideal types, or "ideal workers." Visibly, there is the well-paid unattached, or detachable, worker. These tend to be young men (and sometimes women), who are able to adjust their lives to FIFO work and who can withstand the long hours, or slightly older, professionalized men (and, more rarely, women), who are very highly paid, often engineers with decades of experience in mining operations. Management, for example, is almost exclusively white, imported from the mining company's international operations or recruited from southern Canada.

The worker identified as "low-skilled" and destined for precarious employment is not a failure for the diamond-mining regime, but rather is equally, though differently, desirable. Here, the work of Melissa Wright (1999) is instructive. Discussing export-processing *maquila* factories in northern Mexico, she argues that *maquiladoras* operate through a two-tiered classification of workers as "trainable" or "untrainable":

> The principal marker of the untrainable subject is femininity. As feminist histories of industrialization have noted, the notion of women's untrainability has a genealogy that reaches far beyond the maquila industry. The specificities of this untrainable condition vary depending on how the relations of gender unfold within the matrices of other hierarchical relations found within the workplace: the family, heterosexuality, race, and age, to name but a few. (1999: 463)

In Wright's characterization, the "trainable" worker is a worker in whom investments should be made. This worker is approached as a long-term, stable employee, whereas the "untrainable" worker is a temporary, and potentially disposable, worker who should be used for low-skilled, low-paid, insecure labour. Wright argues that the *maquila* industry marks Latina women as "untrainable," based largely on their roles in biological and social reproduction, and thereby exploits them with greater intensity than their male counterparts. This is a rhetoric mobilized to justify its own discriminatory practices: ones that punish women's reproductive capacities and make *maquiladora* work inaccessible to pregnant women. Wright reports: "'This is not a place for pregnant women,' one supervisor in a machine shop told me. 'They take too many restroom breaks, and then they're gone for a month. It slows us down'" (467). Similarly, in the diamond mines, it is not women's household and community responsibilities that make mining work inaccessible to women, but rather the structures (in particular, the long, inflexible shifts) of the mining regime. In the *maquiladoras*, characterizing women workers as "untrainable" frees employers from having to train them, pay them higher wages, and offer benefits and protections.

Some of Wright's characterizations are not generalizable to the diamond-mining regime. The diamond mines are distinct from the *maquiladoras* – and many other industrial regimes globally – in that they make attempts (however marginally successful) to incorporate women, especially Indigenous women, into work designated as "high-skilled." For example, of the northern women interviewed who had worked for the mines, three (one Indigenous and two non-Indigenous) had received significant training through their mine employment, and while only one had remained working at the diamond mines, all three felt that the mine training had provided a unique opportunity to build a career. Furthermore, given the limited number of northern residents as well as the diamond companies' desire for a local workforce, the level of "disposability" Wright describes in the *maquiladoras* cannot accurately be applied to the diamond industry. For example, while a common theme among women who had worked in the mines was having to leave mine jobs because of their incompatibility with care responsibilities, two women who had worked in admin positions described situations where the mining company moved their work location from camp to town to accommodate their family's needs. That said, Wright's conception of the (extractable and exploitable) value of workers marked as temporary and therefore "untrainable" helps explain the capitalist imperatives for a temporary labour force, the ways in which gender ideology can justify and facilitate this temporariness, and the potential gender and racial violence inherent in temporariness or disposability in an employment regime.[15] In the diamond mines, for jobs labelled as low skilled, such as housekeeping, the primary criterion for employment is flexibility. The work is not desirable, nor, arguably, is it safe (the intensity of sexual harassment of women performing housekeeping work at the mines is discussed below), and it requires a detachment from community and household social reproduction without sufficient compensation for workers to pay someone to take on responsibilities they cannot attend to while at the mine site. Doris, for example, quit her housekeeping job in part because her wage was not sufficient to pay someone to look after her children while she was at the mine site (Interview 111).

For these reasons, women with greater levels of material insecurity have been recruited for these positions. In practice, housekeeping jobs have fallen almost exclusively to local Indigenous women. For example, Arlene Hache, founder of the Centre for Northern Families, said in an interview that in the early days of the diamond mines, companies would call her to see if women accessing the centre would like to work in housekeeping. The Centre for Northern Families is a women's shelter that houses street-affected women, most of them Indigenous. Hache said:

> We would say, "We've got seven women that are interested in the job. Can you hire them for this rotation?" So it used to be based on rotation. And they would say, "Yeah, we can take seven women. Send them up." And those mining people used to say that these women were the hardest workers they'd ever seen. Really hard workers, really thorough. Knew their work, did really well. When they came back [from the rotation], all of that money was blown ... and they could not sustain their sobriety past that ... But, again, these women were not living with partners. They were living with a lot of street violence with partners who were on the street. So it also freed them from that violence. (Interview 110)

Hache's observation that the mines offered a reprieve from street violence for some women is an important one. However, of the women interviewed who had worked at the diamond mine, most – and *all* of the seven Indigenous women and women of colour interviewed who had worked at the diamond mines – had experienced some form of embodied violence at the mine site.[16] This theme was also articulated by participants, mostly Indigenous women, in all three focus groups, though not quantified.

Experiences of violence (in the form of sexual harassment and discrimination at the mine site) are embodied examples of gendered and racialized inequalities exacerbated by the diamond-mining regime. Women who had worked at the mines identified sexual harassment at the mine site as the principal source of personal pain and hardship there. Sexual harassment led to feelings of isolation, lack of safety, pain, and anxiety. For example, Asha explained to me that in her first weeks working in administration at the mines, she wore skirts. She called this "a big mistake" (Interview 118). Since that time, she had worn baggy clothing and isolated herself to avoid sexual harassment. In describing her schedule, she explained that she wakes up at 4:30 a.m. to exercise so as to avoid being sexualized by her co-workers at the gym, and spends her evenings alone in her room:

> I eat in my room every night … I work in the office part so I don't have to interact with certain people if I don't want to. I don't eat in the lunchroom because every time I have, I just get hit on so much and I don't like that. And even one of my supervisors, he says, "Hello, dear sweet thing," and another one, "Hey, good looking" … There's so much sexism, it's disgusting. (Interview 118)

Comments and experiences like this were expressed at their greatest extremes and with the most frequency in relation to jobs in housekeeping. For example, when Doris was hired as a housekeeper, she did not stay in the job long because of the work environment at camp. She felt disrespected by her supervisors and by the way housekeeping staff – all women, mostly Indigenous women – were treated by the male workers. Housekeeping is subcontracted and so is not afforded the labour protections that "real" mine work is given, and housekeeping workers are paid significantly less than most people at camp. Furthermore, the imaginary of sex work shapes the social conditions of work and home for housekeeping staff. At work, sexual harassment is permissible. At home, many of these women experience social shaming rooted in assumptions about what goes on at camp (Interview 202). Indeed, community rumours were rampant: rumours of housekeeping staff engaging in sex work were raised at the two talking circles and the focus group as well as in a number of interviews. For example, at the community workers' focus group, a community service provider said, "I heard they had to take out the ATMs at camp because the men were taking out too much cash for gambling and sexual favours from the housekeepers" (Focus Group One). Doris, speaking as someone who had experienced sexual harassment as a housekeeper, addressed the rumours of sex work:

> You know what? I found that as a housekeeper, the guys always kind of give you hints. Like, "You want to come over?" Like, housekeepers are

good to sleep with or easy to sleep with. You know, like, that's the kind of feeling that you get ... I got a lot of that and I got mad. And one time, I told him, because he's married and he has two small kids, I told him, "You know, if I do that to every guy that asks me, I would have covered maybe half the mine." It just gives you a sick feeling. Because a lot of those guys who do that, they're in a relationship. (Interview 111)

Some women cited unsafe and violent working conditions as a result of sexual harassment as the reason they stopped working at the mines. For others, it was the – related, but distinct – experience of discrimination and harassment. Clarice, a Dene woman who was in a trades training program at the mine, explained that her co-workers resented her success in the program because she was an Indigenous woman. This led to extreme levels of harassment:

C: Yeah. They didn't like that I was climbing this ladder.
R: How could you tell they didn't like it?
C: Because they started acting funny towards me. Like, they would, towards the end it just got worse. Where they were just talking about me and they would do things to the point where it just got really bad. Where we had to all sit down in this room and there were all these men against me.
R: How did you sit down in the room? Did someone mediate it?
C: No. There was nobody that helped. And I didn't know that the union, at the time, could have been there. I didn't know the ropes and stuff ... Those two weeks that I had [off], I was drinking. I was drinking, I was smoking pot ... Just to relax. Just to not think of it. I didn't realize how much it took a toll on me until it was too late. (Interview 103)

Clarice explained that she began to drink and smoke marijuana to relax after the two weeks of harassment at camp, but that the drinking and smoking led to her missing flights and shifts, which further increased the harassment and shame at work. Eventually, she said, "there were times when I just didn't want to live anymore" (Interview 103). Clarice's experience led to her taking a medical leave of absence so that she could seek addictions support and take anxiety medication.

This violence was exacerbated by a lack of social supports as well as by the gendered and racialized colonial ideologies that enabled both harmful acts and impunity for those acts. A few participants explained that, as Indigenous women working in lower-wage work at the mines, they felt they had nowhere to go with their complaints. While working as a housekeeper, Iris spoke of the racist assumptions management made about her because she is Inuk. She recounted a time when she hit her head on a broken cabinet at camp and sought out medical attention:

> They broke one of the cabinets and the door is really heavy and I tried to open it and it landed on my head. I've got the scar from it. And those people that work at the mine, they told me I had to go for a blood test for drugs. I said, "You're not going to look at my head?" And they said, "Oh no, we're going to see if you were doing drugs" ... And here I am, bleeding. (Interview 101)

For Iris, the racism and lack of care displayed when she hit her head was experienced in continuity with violence from other times in her life that had been met with impunity. She saw it as another example in her life of disregard for her health and well-being. These specific experiences of embodied violence must be understood in the context of a gendered and racialized hierarchical employment model, one wherein Indigenous women are often employed in relatively lower-wage precarious jobs wherein sexual harassment is permissible and avenues for support are hard to come by. In the diamond-mining regime, where the temporal imperatives of capital are intensified through the temporariness of extraction and the FIFO tempo, Indigenous women – often made responsible for child care and therefore less available for long-term work – tend to occupy the most temporary of employment locations. Research participants, many of whom moved in and out of mine employment, expressed the violence of this location, experienced as it was at the site of tension between departments of production in the mixed economy.

Conclusion

In this chapter, I have discussed employment in the diamond mines, the first of the three departments of production to be analysed, and the experiences of Indigenous women navigating these structures. While actual employment numbers in the diamond mines are not as high as one might imagine, the diamond mines have imposed an affective need for extraction and have also exacerbated existing socio-economic inequality and precariousness in the region of study. The FIFO regime is a form of capital accumulation that is both spatially fixed and temporally fleeting, a contradiction that plays out in the NWT against the larger racialized and gendered tension between subsistence and capitalist production. The FIFO mine structure requires a detachability from labour

outside of the mine site: specifically, a detachability from social reproduction and subsistence labours, and community and kin relations.
This detachability is made possible by an intensified feminization of
social reproduction and community and household responsibilities, as
well as a racialized marginalization of subsistence, discussed in the following chapters. As a result, while there is a desire on the part of capital
and the state to recruit and retain Indigenous women as mine workers,
Indigenous women have experienced a number of difficulties in accessing and retaining mine employment, due to the incompatibility of the
work with their other non-waged work responsibilities. Indigenous
women have also experienced discriminatory behaviour and attitudes
and violence at camp, examples of embodied and interpersonal manifestations of the structural violence through which the diamond mines
have developed.

7 Social Reproduction and the Diamond-Mining Regime

[Diamonds] are said to be a girl's best friend. I'm not sure which girls they are because it's certainly not anyone in here.

– Talking Circle One

The diamond-mining regime was established in the context of rapid, recent, and profound shifts in social reproduction within northern Dene, Métis, and Inuit communities. The ruptures in day-to-day social reproduction as a result of the diamond-mining regime have been shaped – indeed, enabled – by the longer-standing processes of Canadian state interventions into intergenerational community and household social reproduction. Recall the state programs intervening in Indigenous social reproduction discussed in chapter 3, such as tied welfare payments, the forced resettlement of Indigenous peoples into permanent communities, mandatory schooling, and child apprehension. These are settler colonial initiatives aimed at disciplining Indigenous populations, instituting Western norms, severing the links of intergenerational Indigenous knowledge sharing, and training potential capitalist workers.

As a result of this rapid and violent restructuring, Indigenous households have experienced the boom of diamonds through a new proximity to both the Canadian state and capitalist relations of production. As a grounded way of introducing social reproduction as the site of analysis of this chapter – and its racialized and gendered specificities – I begin by recounting an experience shared with me by Doris, a Dene woman living in Yellowknife. In chapter 1, I introduced Doris, and characterized her engagement in subsistence – specifically, beading and sewing – as an act of resistance, a day-to-day example of prioritizing relations to land and community. Here, I share a more fulsome account of her experience with the diamond-mining regime, for it is through the relational totality of her labours that one can grasp, on the one hand, the interconnected structural and embodied violence of social restructuring as a result of the diamond mines, and, on the other hand, the ways in which the daily labours of Indigenous women can resist the totalizing impulses of capital and the settler state. Doris's story – a story of both violence and resistance – exemplifies social reproduction as a site of place-based subsistence relations in the mixed economy, as well as the temporal emphasis on the accumulation of capital in the diamond-mining regime.

When I asked Doris about the diamond mines, she told me that her ex-husband worked for the mines when they first opened, back when they were married. While Norman was away for his two-week shifts at the mine, Doris took care of the home and their young children on her own. Alongside the day-to-day activities integral to the social reproduction of their household, she engaged in beadwork and sewing, both to fulfil the immediate needs and well-being of family and kin networks, and for sale to fill in the gaps left by Norman's paycheque. However, their new FIFO routine was soon interrupted. In the early years of his mining employment, Norman left Doris for a woman with whom he had been having an affair at the mining camp. He moved out of the home they had just bought, and Doris took on full-time care of their children. Norman told Doris that he could not afford to pay their household bills. Except for some money she made through her craftwork, Doris had no income at the time, and she was not eligible for social assistance, for they were still technically married.

Doris soon found herself unable to pay the household expenses. In a land where temperatures regularly dip below –40°C over the winter, fuel costs are a major expense, and a vital one. That winter, Doris had to pay $500 every two weeks to keep her house warm, a cost she was unable to sustain:

> By that time we had run out of fuel for four days and the house is just solid cold. It was –40 something outside … I had a woodstove, but I had no wood. I didn't know who to ask to go get wood for me. So the only thing I could do is, I told my kids that we were going to be staying in my room now. So I had a little house heater, just a little tiny one. So I covered the door with cloth and then I had that little heater running.
>
> And then my sister comes in the front door...And she said, "Hello? Hello? Is there anybody alive in here?" And she comes in and she nearly flipped out because the floor was just solid frozen. So cold. And I told her, "Yeah, we're in here." "I don't know why you were living like this. You

could have told somebody." And she started crying. And then she left …
She chased her kid to [town] and told him to buy me fuel right now. And
then she went to the grocery store and bought me a whole bunch of bags
of groceries. (Interview 111)

Doris managed to sustain herself and her children that winter with the
help of her extended family. She described the next few years as trau-
matic. When I asked her what had been most helpful for her in this
time, she replied, "My family. Mostly my family. Like when I was struggling so bad, they would help me out here and there. I've always been a
self-supporting person. So I didn't like to ask" (Interview 111).

In the years that followed, Doris faced extreme financial and emotional duress, as well as the court system. Because the state deemed her
home life unstable as a result of poverty, her children were placed in
foster care, in continuity with the high rates of apprehension in Indigenous homes. During this time, Doris took a number of steps to rebuild a
stable home, including pursuing education and training programs and
a variety of jobs. She ended up working as a housekeeper up at the
mines, but left the job quickly, in large part because of the sexual harassment. Furthermore, the FIFO schedule was incompatible with Doris's
responsibilities for her household and extended kin. By the time she
was hired by a mine company, her children were teenagers, and one of
them had a child of their own, for whom Doris was a primary caregiver.
Because housekeeping salaries are significantly lower than the average
salary at the diamond mines, Doris could not afford to pay someone
to look after her children, and she was concerned for their well-being
when she was at camp. So she left her job at the mine to try to make
a living through sewing and beading, craftwork she had learned as a
child and at which she was expert. This was the work she was doing
when she told me her story. Thus, ultimately, Doris was able to rely on
her traditional knowledge and skills to meet her daily needs and the
needs of those who depended on her, thereby divorcing herself from
the extractive regime.

Doris's story touches on many of the common experiences described
in interviews and talking circles by Indigenous women whose partners
worked for a diamond mine. Especially in talking circles, research participants shared their own stories of relationship dissolution resulting
from time away at camp, and the stories of friends and family. As with
other research participants, Doris's experience of the intensification of
feminized responsibilities for social reproduction as a result of the diamond mines was grafted onto newly marketized structures of social
reproduction, unequal gendered access to resources, and the racialized

marginalization of subsistence. These processes are in continuity with – and, indeed, made possible by – the mid-twentieth-century processes of state restructuring of social reproduction that assumed and promoted nuclear structures and feminized structures of care, and in *discontinuity* with the fluidity and interdependence of gendered divisions of labour oriented toward subsistence.

As a result of the heavy demands of Norman's employment in the mines – namely, the FIFO work schedule – Doris's household labour was reoriented toward socially reproducing the capitalist wage labourer: that is, Norman. As this reorientation took place, a new household-level "need" for wage labour emerged as their day-to-day resources were marketized – most notably when they took on a mortgage. As we will see, mortgages and other forms of debt tie people to work at the mines. With Norman at the mines, the household labour performed by Doris was both assumed and subsumed. The separation between (capitalist) production and social reproduction, which is not total in mixed economies, became both an ideological construct – that is, Norman came to be cast as the "breadwinner" and Doris as the "caregiver," regardless of the forms of labour in which she engaged – and a temporal, spatial, and material reality, with Norman spending more than half of his time 270 kilometres away from their shared home. For Doris, there was a violence to this shift, both structural and embodied (environmental, financial, sexual, and physical).[1] While Doris's kin helped her keep afloat, the isolation she described in recounting the winter Norman left is telling. Doris's family helped by providing food and firewood, but Doris was also struggling with the new demands of a mortgage, a household-level responsibility that is not easily accommodated by inter-household social relations.

Throughout, Doris engaged in subsistence production and distribution networks to manage the ruptures in her life. The role of subsistence is complex here, in part because of the mutability between social reproduction and subsistence in Dene modes of life. Doris engaged in beadwork and sewing for different purposes and in different relationships to capital, from immediate consumption to informal sale, and in the end, she developed her own business. However, a continuity persisted through Doris's labour over time, related to the subjective and place-based meaning she ascribed to beadwork and sewing, rooted as it was in the knowledge and traditions of her kin and community. Belying the settler colonial history of devaluing Indigenous women's subsistence labour, ultimately it was Doris's expertise in beadwork and sewing that freed her from the need for employment in the extractive regime. Most of the stories that women shared did not conclude with such a clear shift toward subsistence. However, the general experience of rupture and resistance at the site of social reproduction, and the choices Indigenous women made to reproduce the social relations of their community, were common and powerful themes in interviews. The present chapter focuses on the characteristics of this rupture and resistance at the site of day-to-day social reproduction.

When we locate our analysis of the impact of diamond mining outside the physical sites of extraction, the violent processes of restructuring on which the diamond mining regime is predicated become clear, both materially (i.e., in terms of access to resources or lack thereof, corporeal violence, and environmental destruction) and ideologically (i.e., in terms of processes of racialization of Dene, Métis, and Inuit people that justify and perpetuate new accumulation and restructuring for the pursuit of resource extraction, and the intensification of a Western male-breadwinner/female-caregiver model). This restructuring is a dialectical process: a contested and uneven movement responding to the demands of capital in the form of diamond mining, wherein social reproduction is the locus of struggle between settler capitalist and Indigenous materialities and ideologies of being and labouring. Thus, rather than casting the extractive regime as the agent of change and Indigenous women as the bodies on which change is enacted (passive victims, or recipients, of a colonial process), I emphasize the agency and power inherent in the stories shared by northern women who have shaped, and are shaped by, the material and ideological structures described herein.

In the two sections of this chapter, I characterize the rupture at the site of the social reproduction performed by Indigenous women under the diamond-mining regime as a reorientation (Section I) and a restructuring (Section II). In Section I, I argue that the intensified engagement with the diamond-mining regime and the particular demands of the FIFO tempo have resulted in a *reorientation* of social reproduction toward capitalist production and away from subsistence. This reorientation emphasizes the nuclear family to the detriment of inter-household relations. In Section II, I argue that this reorientation has been accompanied by a *restructuring* of social reproduction as a result of the spatial, material, and ideological separation of capitalist production from social reproduction, one that relies on and perpetuates a Western male-breadwinner/female-caregiver model. Both of these processes, and particularly the attempted separation of capitalist production from social reproduction, are facilitated through an (again, incomplete and divergent) redefinition in gender roles, including a racialized devaluation of the labour involved in the daily and intergenerational social reproduction of the place-based relations of the mixed economy. Drawing on research participants' narratives, I discuss the diverse ways in which Indigenous women have negotiated this rupture and resisted new settler colonial imperatives while also attending to the structural and embodied violence woven into these processes that targets Indigenous women.

This chapter focuses on day-to-day processes of social reproduction; the following chapter analyses the intergenerational community-level reproduction of subsistence relations in the regional mixed economy. These are not discrete subjects, in that both chapters are concerned with the tension between place-based and temporal modes of production; however, this chapter focuses on the FIFO tempo and its impact on day-to-day social reproduction, whereas the next chapter emphasizes the diamond-mining regime's temporariness and its impact on intergenerational social reproduction.

I. Reorienting Social Reproduction

The impact of the diamond mines on relations of social reproduction in the Wıı̀lıı̀deh region plays out through the persistent tension between place-based subsistence relations and time-based capitalist relations. For some – namely, settlers who had relied on the gold-mining regime for wage labour – the diamond mines represented a welcome continuity: a reinstalment of the "temporary permanence" of high-wage extraction, albeit through a new work tempo and an intensified temporariness. Particularly because the gold mines had ended so violently and abruptly, for some parts of the Yellowknife community the diamond mines were a way of resolving an earlier rupture and maintaining an extractive political economy. For example, Jenny's personal experience working at the diamond mine was characterized by a sense of alienation, which she felt was largely the result of a masculinized workspace, yet she also noted the ways in which diamond companies had contributed to community infrastructure and resources in the NWT:

> Eventually I was [working for an NGO]. And I think it was Ekati, they gave us $10,000 of multi-year funding, which was great. So $10,000 every year to do youth workshops during our festival. So that's what it is, right? They have these large amounts of cash that they can drop on things. And make things happen.
>
> And that's actually very in line with the whole trajectory of how Yellowknife came to be in the first place. If you look at Stanton Territorial Hospital and some of the sporting infrastructure, [it] came about through Con Mine and Giant Mine. Those direct infusions of large amounts of

cash, which have absolutely benefited the community of Yellowknife ... Who knows, but for sure, in this place, it's definitely made a huge mark. (Interview 115)

Jenny's observation illustrates the discrepancy in the effect that the diamond-mining regime has had on social reproduction in the area around Yellowknife. At the community level, for some local organizations, networks, and people, the diamond mines maintained the continuity of a local socio-economy built (albeit in fits and starts) by extractive dollars. In interviews and focus groups, community workers noted the inconsistency of funding from the diamond mines for their organizational activities (see below), but most were also quick to note the ways in which money from the mining companies had contributed to their programs. And certainly, some interviewees – most but not all of them non-Indigenous – mentioned feeling relieved that the diamond mines were providing jobs. Heather, for example, described the enthusiasm with which her former partner and his friends, who had performed seasonal work for the gold mines, initially took up training and job opportunities with the diamond mines.

The continuity in the "temporary permanence" of settler extraction contrasts with the tension at the level of day-to-day social reproduction of Indigenous households and communities as a result of the shift toward the imperatives of the diamond-mining regime. As the first extractive regime to engage the region's communities in large-scale, sustained employment, the diamond-mining regime exerted a new pressure on social reproduction in households and on the mixed economy more generally. Notwithstanding their diverse social locations and employment histories and experiences, all of the interviewees and talking circle participants articulated their experiences and observations of the diamond mines as a major upheaval in local relations among humans, non-human animals, and the land. This evokes the Tłı̨chǫ concept of ruptures in social relations that must be negotiated and resolved – ruptures that interrupted processes of social reproduction at both an interpersonal and community level. Dana, a community worker, recalled the experiences of the households with which she worked during the early years of the diamond-mine regime: "The well-being of the family was in disarray. Because this was a time when you got an income, you're going to make a life that's going to be good, and yet everything is turned upside down. And it's the women, the women and the children, that have the worst part of this craziness" (Talking Circle One).

While rupture involves challenges, and possibly violence, recall that it does not solely denote harm. Given the high cost of living and the limited government provision of day-to-day needs, which had been clawed back under neoliberalism, a significant number of research participants pointed to the benefits that a high income could offer a family. Drawing on their own lives or those of friends and family, some of them gave examples of increased material stability and access to resources and experiences that had previously been unavailable: things like family holidays away from the NWT and lessons and activities for children. As Madeline put it, describing her observations of Indigenous families in which someone was employed by the diamond mines:

> The money was definitely a very positive thing. Oftentimes, it pulled them out of poverty. It was a matter of poverty or a somewhat good standard of living. You know, providing for the kids. Maybe having a vacation. Maybe having a vehicle. These were all things that were not accessible without that mine job. You know, the level of income jump is just so drastic. (Interview 302)

Work at the diamond mines provided very real material benefits to some. So it is notable that, when those benefits were raised in the two talking circles and the focus group, the research participants expressed a consensus that the financial benefits of the diamond mines came with caveats.[2] Participants distinguished between the material rewards of high-wage work and its impact on emotional or social well-being, noting that the former do not secure the latter. For some, an increased income as a result of the diamond mines simply amplified existing household dynamics, whatever they were. For others, the increased income resulted in new inequalities, new and unwelcome capitalist imperatives in the community and in households, and the possibility of intensified embodied violence. Most participants commented on the reorientation of social reproduction toward the demands of the diamond mines and how this touched both the community and individual households.

Indeed, as a result of the new domination of this particular form of wage labour, many Indigenous women in the interviews and talking circles recounted pressures to reorient their day-to-day social reproduction toward the demands of capital, potentially displacing community resources and interrupting land and community relations. Beth, a Dene woman living in Behchokǫ̀, put it this way:

> It's different now from way back when I was younger. It's really different. Because back then everyone was always in the community. There were always things happening. And people, they were like a community. And right now it kind of seems like everyone's separated. And it's not a

community, the way it used to be when we were younger. We had things to do. We had a sports complex where we could play sports. They still have the occasional summer baseball, kids play in the basketball court there. They go swimming at the T-bridge. That's still happening.

But back when we were younger, we had more places to go. Like, there was a restaurant there. We used to have a restaurant here. Just right across behind the firehall, across the street. There was like an arcade-restaurant, they had games and pool tables and they served food. And that was our main hangout. We would either hang out there or else go to the sportsplex. Those would be the two main areas. And then we'd just walk around here, this whole area, this one little loop. You would just see people wandering around. Like, I don't know how many times we would go around the loop. Like, during the evenings. You'd go with friends and talk. Or meet up with other people. But right now it's really different. (Interview 107)

Beth experienced the loss of community resources in actual bodies: the bodies away from the community due to FIFO work and the requisite time away from the community for diamond-mine employees.

Even when mine employees were back in the community, many employees and their families faced constraints on their social reproduction that began to orient their labour toward capital. The move from public (government-subsidized) housing to private housing – a marketization of household resources – is a particularly damaging example of this shift in orientation, and many research participants raised it as an issue. Recall that around Yellowknife, the development of the diamond mines was accompanied by a decline in public housing through tightened criteria for tenants and by an increase, in both price and quantity, of market housing. This, combined with a significant jump in household income, meant that newly employed workers in the diamond mines were navigating either significantly higher rents for public housing or their first experience with market rents or mortgages. For those Indigenous households for whom engagement with the diamond-mining regime marked a move from public housing to market rents, a new "need" for wage labour formed as their daily household needs were marketized – most notably, when they took on a mortgage. Particularly in the early days of the mines, workers were targeted by banks as potential new homeowners and encouraged to financialize their assets through financial literacy training sponsored by the diamond mines themselves (Focus Group One). Mortgages and other forms of debt tied people to mine work; thus, debt was mobilized as a means of dispossession (Soederberg 2014), further constraining the capacity of Indigenous people to orient their labour toward subsistence and the reproduction of the mixed economy. Clarice, who had shared her experience of harassment by her all-male co-workers (see chapter 6), described the constraints of debt in this way:

> I wouldn't mind a change of career. It's just that I, in the process [of working in the diamond mines], I ended up buying a trailer. And I have a vehicle and all these bills. So it's kind of like I'm stuck there, unless I find something really similar to the pay. Because the pay is good. But I shouldn't let it affect my health. Like I said, it's like a touch and go thing. It's an in-between thing. It's like you don't have a choice. (Interview 103)

For many research participants, the experience of feeling bound to wage labour by debt was new. For significant numbers of northern Indigenous people across the NWT, public housing is the only form of housing they have known since being relocated to permanent settlements. In public housing, rent is collected on a sliding scale determined by household income. Through the NWT Housing Commission, subsidized housing is offered so that, say, a person with an income of less than $1,667 will pay $70/month in rent, while someone with an income higher than $8,334 per month will receive no subsidy and will pay up to $1,625 a month (NWT Housing Commission 2015). In many small communities, government housing is the only housing there is. Sliding-scale rent payments make it possible for people to meet their needs in mixed economies oriented toward subsistence, as consistent, high wages are not required for access to housing. Thus, while public housing emerged over the course of the settler colonial drive to move Indigenous peoples into sedentary settlements, contemporaneously, it also plays a role in socially reproducing the mixed economy. When someone in a household is working at the diamond mines, that household no longer qualifies for reduced rent in public housing. Many families remain in public housing, particularly in Behchokǫ̀, N'dilo, and Dettah, where market rentals are virtually unavailable. This means that for families who remain in public housing, their quality of life stays the same while their rent skyrockets. Arlene Hache offered an example:

> This one family ... This guy just wouldn't pay his rent. He was in government housing, again. So you go for years not paying rent, or paying totally minimal rent, to paying massive amounts of rent ... So we're not talking about huge amounts of rent for a house ... Their rent goes from $32 a month to, like, $2,400 a month for the same shack. That's not even like, "I moved into a three-bedroom house. I have to pay more." It's the same shack they've lived in for the last twenty years and all of a sudden they

have to pay massive amounts of rent for that. So he didn't pay that. So eventually he would get into trouble and she would go to income support and say, "Help me pay my rent." And they're looking at all this money coming into the family, saying that's not going to happen. (Interview 110)

Other research participants recounted that they had been able to move into a nicer home as a result of the diamond mines (Interview 106), or to relocate to Yellowknife so that they could have greater access to municipal resources and more choice in schools and activities for their children (Interview 102). However, Arlene's narrative speaks to the pressure generated when housing requirements bump against high wages. A significant number of people with whom I spoke made note of this shock in their own households and in their community at large. And loans were being taken out not just for housing: many participants discussed the "big toys" that came with the diamond mines – specifically, trucks and boats. Ford opened a dealership in Behchokǫ̀ shortly after the first mine opened, catering to community members newly employed by the mines (Interviews 110, 303). A number of participants, particularly those living in N'dilo, Dettah, and Behchokǫ̀, or those from smaller communities living in Yellowknife, described the community-level tensions and resentments that developed in response to the "big toy" culture, which put on sharp display the new inequalities and fomented polarization within communities.

Mortgages and other forms of debt tie people and households to wage labour in new ways, and in addition to that, buying a home or switching to market rent in public housing can introduce new insecurities, stresses, and inequalities within and between households. Arlene told another story about a family with whom she had worked that illustrated the particular experience of entering market-oriented housing in a place like Yellowknife, where the cost of living is high and the consequences of being unable to pay for services like utilities are dire, because of the cold:

The other thing I thought about is this family of ten kids I know where they have a ton of money come into their house and they got a mortgage from the GNWT to get a house in Yellowknife. Which is very expensive ... They were living in a very small community. So they lived in a tiny community. I think the couple lived with his family, so they hadn't paid rent ever. And I don't even think his family had paid rent, really, because it was government housing. So all of a sudden he comes to Yellowknife and he has a big job and he bought a house with this mortgage. Tons of utilities because utilities are through the roof. Lots of new furniture. Like, he really stocked up.

And eventually, I mean, it's good for the first while because everything works. But ... they just got into a horrible situation with their bank. And ... the woman was running around with ten kids not really knowing what to do. And her house froze and it was horrible. Because essentially they lived in an unheated house with water pipes busted for months and months and months. Put themselves at risk health-wise. And their family at risk. And their kids at risk. With no real resource to figure it out.

And nobody felt a great deal of sympathy for them because they were getting a huge amount of money into the house [from his employment with the diamond mines]. So everybody would say, "Well, I'm not helping you." So, anyway, this couple did try to do a heck of a lot to mitigate everything and in the end, I think they slowly dug their way out under tremendous stress. (Interview 110)

Arlene's story speaks to the experience of reoriented social reproduction in a number of ways. It is an example of the migrations from small communities to Yellowknife, which often result in the marketization of not just housing but food and household goods as well. Having moved to Yellowknife from a smaller community, the family in Arlene's story found itself separated from kin and community networks at a time when they were struggling to manage new financial requirements for their daily maintenance. For those who have left community networks behind, an inability to pay rent can have severe consequences. Small communities in the North offer resources tied to the land as well as interdependent networks of care and distribution (Dombrowski et al. 2013); conversely, when social reproduction is marketized, the responsibility tends to fall squarely on the nuclear household. This tendency is amplified and spatialized when families move to Yellowknife, away from their home communities. One participant suggested that some of the homeless in Yellowknife were people who worked in the diamond mines who had abruptly lost their jobs (Interview 109). While this direct link could not be substantiated, the observation points to the loss of the security of kin and community networks through the move from smaller communities to Yellowknife, in the context of the higher cost of living and the reduced availability of public housing (Christensen 2017). At the same time, the smaller communities lost people to Yellowknife, a concern outlined by Beth. Certainly, moving to Yellowknife is an understandable choice – for some, it reduces the commute time from the diamond mines; for others, Yellowknife's allure has more to do with the resources of an urban centre, including more schools, social services, and activities. However, the move from smaller communities to Yellowknife is a reminder that the reorientation of day-to-day social reproduction toward the demands of capitalist production is accompanied by rifts in intergenerational community social reproduction (see the following chapter).

The family Arlene described had previously not required a steady high income to meet their daily needs. Rather, like many households in the territory, they had been accustomed to meeting those needs through a combination of wage labour, subsistence, interdependent supports from kin and community, and (most likely) treaty payments. Many participants described the new, or deepened, need for wages as a source of stress, tension, insecurity, and/or embodied violence – characterizations rooted in part in the instability of diamond mine employment, but also in the new power of wage labour within the triad of social reproduction, capitalist production, and subsistence. This insecurity is gendered, both because of the different roles women and men in a household might take up in times of financial crisis[3] and because of gendered inequality in responsibility for social reproduction and in access to diamond mine employment and household finances.

Research participants noted with concern that most diamond-mine workers are men and that in many households it is men who control the diamond mine salary. This concern was raised in all three focus groups and in a number of interviews. In Yellowknife and Behchokǫ̀, where most Indigenous diamond-mine workers have been men, and where Indigenous women are more likely to find employment in government offices (municipal, territorial, and federal in Yellowknife, national [Tłı̨chǫ Nation] in Behchokǫ̀), the effect of the diamond mines has been less a greater accessibility to wage labour overall, than a gendered shift in access. For some Indigenous women, this has resulted in a new household-level material inequality; for others, the primary impact has been increased responsibilities for social reproduction alongside their wage labour. Indeed, a number of women interviewed were workers in government or government-related offices who had been ideologically and materially reconstituted as the primary caregivers in their households as a result of the diamond mines. That shift in responsibilities was not accompanied by any actual change in their employment.

Many research participants expressed concern about the implications of unequal access to household money for women with partners who use violence. This theme was particularly strong in the community worker focus group and among interviewees, but it was also expressed in the community talking circles and in interviews with women who had direct relationships to the diamond mines, sometimes in relation to participants' own stories and sometimes in discussions of their friends and family. In the Yellowknife region, whether social services are available or not, the small populations and distances between towns make leaving a physically violent partner difficult, and far more so if financial resources are limited. For a survivor, social services such as separate housing from a partner can become much more difficult to access when that partner works at the diamond mine and the household is no longer designated as low-income. Rose, a young Dene woman living in Yellow-knife, had been in a relationship with a man that became violent after he started working at the diamond mine. In an interview, she explained why she stayed in the relationship: "It was financial. That's what it was. I had a household to support. I needed things for my [child]" (Interview 102). Rose's material insecurity as a result of relying on her partner's wage labour was coupled with taking on sole responsibility for the day-to-day social reproduction of their household (including their infant), a shift related to the new, or newly deepened, separation of social reproduction from production as a result of the diamond mines. Rose explained that, while she had always imagined herself being a working parent, it was initially difficult to enact that role because her partner was away so much of the time. Like Doris, she credited her kin and family networks as the resources that helped her leave her partner. Indeed, community and kin networks emerged as a common theme in discussions about resisting violence and the orientation toward the diamond mines; for Rose and Doris, kin and community networks helped relieve intensified individualized/nuclear responsibilities of care. Inter-household care and distribution networks, a crucial element of subsistence, are discussed in the next chapter.

Community-level spaces and relations were not exempt from the embodied implications of the structural reorientation imposed by the diamond mines. Both community workers and women who self-identified as having been impacted by the diamond mines expressed concerns regarding the relationship between problematic substance use (mainly among men) and violence. Indeed, while comments on problematic substance use were not solicited, concerns about the impact of new street drugs that were arriving with the diamond mines were a common theme in both interviews and focus groups. Eloise, a Dene woman living in Behchokǫ̀, made this observation:

> Drug dealers are coming this way, coming to make money, I guess. Because the diamond mines are here. The money's here. So not only companies are coming up here, and businesses, but drug dealers are coming up here. (Interview 106)

Scholars have noted (Tsetta et al. 2005, Government of Australia 2014) that mine work, in general, and the psychosocial pressures and uneven

schedules of FIFO work, in particular, can contribute to substance mis-
use, depression, and violence. Such problems would have major social
implications anywhere. However, in and around Yellowknife, residents
experience the psychosocial impacts of the mining regime in the context of the historical and contemporary processes of settler colonialism
and the ruptures resulting from the intensity of capitalist and Canadian
state-driven restructuring since the mid-twentieth century: upheaval
upon upheaval.

Articulating the relationship between problematic substance use and
gender violence is a difficult endeavour, because this relationship is often emphasized in order to individualize responsibility for violence,
usually along racialized lines (as noted by Davis 1981; hooks 1984; True
2012). However, when substance use is approached as part of a social
shift, it is possible to link embodied violence and problematic substance
use as tendencies related to and exacerbated by the diamond-mine regime, without expressing an individualized correlative relationship
between substance use and violence. Research participants linked the
increased street violence in Yellowknife and smaller communities to the
new hard drugs that had come north with the diamond mines. Speaking
about the new presence of drug-related street violence in Yellowknife,
Annie said, "There's times with the snow on the sidewalk, I see blood.
I see blood trailing and people are really violent to each other" (Interview 208). In interviews, people brought up drugs and alcohol most often when asked questions about how diamond mining is related to the
dissolution of relationships and violence against Indigenous women in
northern communities. Elizabeth, for example, gave a detailed account
of her partner's struggle with substance use. He had worked seasonally
for the gold mines and then taken on full-time FIFO work when the diamond mines were established. He drank during the gold-mining era –
a habit she described as harmless – and started using crack cocaine
when he began working for the diamond mines (Interview 104). For
Elizabeth, her partner's problematic substance use was an example of
the new brutality that accompanied the diamond mines. Jenna, a Dene
woman living in Behchokǫ̀, made the following comment:

> Well, again, drugs, right. That's how it happens. There was a lady here
> today who was just blue. She said, "I was dragged out of my house by my
> hair." And I was so sad for her. They don't know how it affected them. Not
> for the guys as well as women. Like in the two weeks off from each other
> and they come back to violence. Drugs and alcohol, you know? They're
> working so hard for twelve hours a day for two weeks and then it's all
> gone. They've got nothing to show for it. It's ever sad. (Interview 109)

The violence associated with substance use is tightly wound with unequal access to financial resources, a characteristic of the region intensified by the diamond mines. However, for most research participants, the diamond mines did far more than magnify existing experiences of embodied violence and substance abuse. A number of participants who spoke about drugs, for example, noted a *qualitative* shift, describing the new forms of embodied violence – both intimate partner violence and street violence – since hard drug dealers came to Yellowknife. It would be inaccurate to reduce this violence simply to a result of the impacts of the diamond mines, given that violence against Indigenous women is a social phenomenon that cannot be reduced to one aspect of social change but rather must be addressed holistically in the context of settler colonialism. That said, the narratives shared by research participants offer grounded insight into the ways in which the structural reorientation of labour toward the demands of capital and away from place-based relations has gendered and violent embodied implications.

III. Restructuring Social Reproduction

Before discussing the separation of processes of social reproduction from capitalist production as it relates to the diamond-mining regime, I review the theory of this process at the general level of the capitalist mode of production to highlight how integral this separation is for the creation of surplus-value and, it follows, capital accumulation. Recall here Marx's two uses of mode of production: as the concrete and technical processes of one particular mode of production – in this case, the northern diamond-mining regime; and as a global stage of development. I have been using the term in line with the former definition, but here I relate the separation of social reproduction and capitalist production to the latter. While capitalist production is materially inseparable from the daily and intergenerational reproduction of labour power (i.e., social reproduction) in that it quite simply cannot function without it, the creation of new (capitalist) value relies on an obfuscation of this relationship. This is because new value, or profit, is created under capitalism by paying workers less than the value that their labour contributes to the commodities they produce. This is, in Marxian political economy, the exploitation of workers.[4] The underpaying of labour is facilitated in part by the racialized invisibilization of subsistence (discussed in the next chapter) and the naturalization of social reproduction – that is, through the ideological conflation of "work" writ large with capitalist production. This narrowing of labour allows capitalists to abstract themselves from the reality of the multiple labours and resources required to reproduce the worker, a necessary abstraction for exploitation. Thus, the creation of new surplus-value requires a division between capitalist production and the non-capitalist forms of labour on which it relies, a division that is both real and abstract. For example, the biological and social reproduction of children is necessary for the reproduction of capitalist relations; however, it is framed as naturalized and feminized care work and as the individual choice and responsibility of families, especially women. This framing allows employers to avoid paying true costs or accommodating parental leaves and the many demands of child care (noting, certainly, that employer and state approaches to child care vary vastly across political-economic conjunctures and space and have been a key site of successful feminist action over time).

This division is abstract in the sense that it describes an ideological division between capitalist (seen as "real") labour and non-capitalist labour imposed on forms of labour that, materially, are interdependent. However, the ideological separation of capitalist production and social reproduction has myriad – and dialectically shifting – material consequences: in this sense, it is a "real abstraction," to use Lucio Coletti's (1972) term. Indeed, accumulation under capitalism relies on the exploitation of workers' unpaid labour to create surplus-value. This exploitation is something that occurs not only through methods of exploitation in the workplace but also through the assumption that workers will be reproduced free of charge for the capitalist. This is why the accumulation of new spaces of capital is always, necessarily, also the disciplining and restructuring of social reproduction.

Silvia Federici (2004) analyses the disciplining of women's bodies and relations of social reproduction during the original capital accumulation in Europe, arguing that "one of the preconditions for capitalist development was the process that Michel Foucault defined as 'disciplining of the body,' which ... consisted of an attempt by state and church to transform the individual's power into labor-power" (133). In line with Marx's vivid descriptions of the violent disciplining of the body (Marx 1976: 896), Federici argues that the creation of The Worker required a regime of terror aimed at relieving the body of its dynamic potential – what Federici calls the body's "pre-capitalist magic" – and turning it into a machine for the accumulation of capital. Federici's (2004) embodied subject is both female and male. However, she argues that there is a particular violence against women and feminized bodies; indeed, "the rise of capitalism was coeval with a war against women" (14). Federici writes that capitalist production is only possible through an ongoing violent appropriation of women's lives and bodies, and she points to the ways in which the state has used gender violence to regulate social reproduction for the demands of capital.[5] Thus, capitalist colonialism, in its various forms, has brought with it a new material imperative for disciplining of, and violence against, women. These processes are made possible through the various racisms produced through processes of colonialism, imperialism, and capitalist development.

Returning to the specifics of the political economy of the region of study, the separation between social reproduction and capitalist production was made possible by the colonial-historical and contemporary processes targeting subsistence practices. These processes embraced the liberal capitalist assumption that (masculinized) production is separate from (feminized) social reproduction, or at least the tacit ideological valuation of this separation. Without the activist role of the Canadian settler state responding to the demands of capital and disciplining the bodies within its borders accordingly, there is no real separation between subsistence and social reproduction in the mixed economy. As we saw in chapter 3, it was racialized restructuring processes that feminized and made distinct social reproduction.

In the NWT mixed economy, the FIFO extractive structure uniquely imposes a spatial expression on the division between capitalist production and social reproduction. This structure is, perhaps, the most readily visible of the extractive regime's characteristics. This is what distinguishes the diamond mines from the gold mines and other earlier NWT extractive projects, with profound implications for the relations between departments of production, as well as for Indigenous women's (potentially violent) experiences of those relations. The assumption that the capitalist wage labourer is an unencumbered individual free of any responsibility for care manifests itself when workers must agree to live away from their home, family, and community for half of their lives in two-week shifts during which they are worked so hard (twelve hours a day for fourteen days straight) that they come home exhausted, in need of rest and replenishment at the precise moment when many of their partners are desperate for a reprieve from – effectively – single parenthood. I use the term "single parenthood" here because participants commonly articulated it in interviews as a way of expressing the extremely imbalanced social reproduction responsibilities imposed by the FIFO structure, as well as the loneliness and isolation experienced by many of them as they performed social reproduction for two-week stretches (and sometimes more) without the companionship and support of their partner.

Of course, workers' day-to-day social reproduction is undertaken by paid staff at the mine camp, but beyond having one's food served in a cafeteria, having one's room cleaned, and enjoying site-specific perks, like a gym or an ice cream bar, the social reproduction of mine workers is entirely externalized. To put it another way, the day-to-day social reproduction of workers is performed on site, but their more complex, interpersonal, community-based and long-term needs are divorced from the work site. For example, workers' responsibilities and contributions to subsistence, and to daily and intergenerational social reproduction in households and communities, are structurally obscured. This separation means that women are made responsible for socially reproducing the household as well as for broader inter-household, community-level social reproduction. The FIFO structure thus facilitates workers as the embodiment of variable capital, as vessels of living labour that can be flown in, drawn upon as much as possible, and, after they have been depleted, flown away for temporary replenishment or until they're needed again.

The other half of the FIFO equation reflects the assumption that day-to-day social reproduction can and will take place, regardless of this disruption, at "home": the site of non-work, restructured toward the very heavy demands of FIFO labour. One of the strongest themes that came out of interviews with Indigenous women was the material burden of day-to-day social reproduction they took on as primary caregivers of children while their partners did FIFO work.[6] Madeline discussed her work with women in the Yellowknife community:

> This is the biggest impact I see on the family unit … These women are single parents … I knew someone would come up to do this research because I feel like I live in a community where families are fragmented on purpose. We choose to remove half of the caregivers half of the time. How can this not have a significant impact on raising a family or being in a marriage? (Interview 302)

All research participants who were women whose partners were working in the diamond mines, and, as a result, were the primary caregivers in their household, expressed some level of fatigue from their daily household labours, ranging from minor frustration to extreme exhaustion. Tina, a Dene woman living in one of the First Nations communities in close proximity to Yellowknife, had worked for well over a decade as the primary caregiver while her partner worked at the diamond mines:

> Yeah, fourteen years and I am just so tired of it. Keeping the kids for myself, and when they're sick, I have to deal with it. And then he misses out on the things that they do. You know what I mean? He missed out on her

graduation, when she graduated from kindergarten. Missing her [sports] tournaments, and speeches that my son just did. He missed that. And my baby misses him. He's constantly asking for him, nightly. It's hard. (Interview 112)

Tina's experience was echoed by a number of women. For example, Rose spoke of her partner's initial employment with the diamond mines:

He went away for his first rotation when my child was three weeks old … It was emotional. Because obviously having a baby in the first place is a huge transition. And then having him go when the baby was three weeks old was terrifying. And being a first-time mom, and being young, I was terrified. (Interview 102)

Beth said much the same:

Because he's out there for the whole two weeks, sometimes my kids would get sick and sometimes I would get sick. And those times would be the most difficult. And a few times I really wanted him home with me to help me … I wanted him to come home. But they said they didn't have enough people up there to take on his role, so he couldn't. (Interview 107)

Notably, while the women we interviewed expected to have twenty-four-hour care responsibility during the two weeks their husbands were away at the diamond mine, most of them also expressed difficulties shifting divisions of household labour once their partners returned. For some, this was expressed simply as a bumpy transition in living arrangements, with shifting household routines, habits, and rules. For example, Rose said:

So it was this constant flip-flop, going back and forth. And you're not used to living with someone. So even him being around, like, I'm used to managing me and our child. Then him coming around and I'm like, "I don't know how to do this. Get your socks out of here." (Interview 102)

For most women, however, the difficulty in shifting toward a more equitable distribution of care work upon their partner's return involved much more than simply adjusting to a new body in the house. Diamond-mine workers tended to return home exhausted, depleted, and in need of rest, and often feeling that they should not have to do work on their "time off." For some of them, bodily and emotional exhaustion was coupled with the ideological justification of "women's work" and "men's work"; as a result, the materiality of women's responsibility for social reproduction did not change significantly when the man was home. As Beth noted, "even when he's home, I still feel like I have to do everything" (Interview 107).

The male-breadwinner/female-caregiver model *de facto* encouraged by the diamond-mining regime intensified the nuclear orientation of both men and women's labour: for men, because they were away from home and community, their "contribution" tended to be financial and thus more easily absorbed directly by their household, rather than shared by the community, as would be the case with subsistence activities. Research participants noted the increased roles and responsibilities Indigenous women took up in the community as a result of men being away at camp; some also discussed the challenges of maintaining strong community relations when they carried their household-level responsibilities on their own. To be sure, this tendency toward making social reproduction "nuclear" was experienced unevenly by research participants, resisted as it was through inter-household kin and community relations. The demonstrable commitment to sustaining and reproducing strong kin and community linkages is an example of the ways in which the rupture of the diamond mines was negotiated through the maintenance of Indigenous modes of life. Specifically, inter-household structures of social reproduction resist the pressure – at work from the mid-twentieth century to today – to adhere to Western nuclear models of social reproduction. Thus, one should read the analysis here with the understanding that women focus on household-level demands not because their understanding of social reproduction stops at the walls of their home, but because their work and responsibilities within the household have intensified.

In interviews with women who worked at the diamond mines, it was clear that their experience – and the experience of their partners – did not correspond with that of men who worked at the diamond mines while women partners were the primary caregivers. Instead, for women research participants who worked at the diamond mines, day-to-day social reproduction continued to be a significant responsibility, and juggling this responsibility alongside wage labour in the mines was a major challenge – one that, for some research participants, meant that they did not last long in mine work. What emerged in interviews with women whose male partners worked at the diamond mines was a pattern of shifting gender roles wherein women were ideologically deemed responsible for social reproduction, often newly concentrated at household levels. Certainly, household gender roles are built through historical and contemporary processes far broader than the diamond mines themselves, and my purpose here is not to suggest a simple causative shift, nor that households had entirely equal or equitable gendered divisions of labour prior to the diamond mines. Instead, I am suggesting that the social effects of the spatiality of the FIFO diamond-mining regime have been bolstered by an intensified male-breadwinner/female-caregiver model supported by the mine camps. FIFO mining camps are notorious breeding grounds for hypermasculinity, as a combined result of their disassociation from the cultural norms in the towns from which the workers come, the predominant presence of men and associations of mining with "masculinity," heavy work demands that can cause exhaustion and stress, and separation from family and loved ones. Indeed, the Inquiry into Missing and Murdered Indigenous Women and Girls in Canada linked increased regional rates of violence against Indigenous women in British Columbia with the presence of what they called "man camps" (MMIWG 2019), discussed further below.[7]

Rose, for her part, noted the impact of the hypermasculine culture at camp on her relations with her former partner at home:

> When he'd come back, I'd ask him simple things to help me around the house and he felt like – and that was the difference between us, our ideas, like gender roles, all that stuff. I thought it was like, "Let's share stuff. We'll both build a household," and he was like, in "You're trying to make me a housewife." (Interview 102)

Rose felt that her partner's attitude was newly developed through his work at the diamond mines. She discussed his shift in beliefs through working at the mine:

> R: I was expecting shared. Like, sharing household duties. I understood that he went away. But my intention was always, after my year of mat leave, I would return to work. So that was really challenging. And then, him. He was young, too. Around my age. And as time passed on, as months and years passed on, I became really resentful of even the men he worked with at the mine. Because I definitely saw a shift ... What was the shift?
> P: ... It was gradual, right? But I always felt, like, it makes me feel, because these are my personal feelings. I really felt like the men he worked with up there were sexist, were racist, were old, were jaded, were bitter ... I think they were racist against all races. Not even just First Nations

people. Because he would come back with these ideas. I'm like, "Are you crazy? You're Aboriginal. What are you doing?"

But I would say they were older white men. Not always. But these were the journeymen when he was an apprentice. They were training him. He looked up to them. And they had twenty, thirty years' experience. And relationships and divorces and all kinds of things. And so I know, even at that time, I was like, "What are you thinking?" From where we were, our ideas when we were pregnant, to him being away literally half the year. (Interview 102)

As Rose articulates, the new gendered expectations her partner learned through his work at the diamond mines were racialized. The – tacitly white – hypermasculinity encouraged by senior mine workers intersected with spoken and unspoken racisms, and the new expectations and judgments her partner brought home clashed with the practices of interdependent gendered household roles she had learned through her Dene kin and community.

The ideological nature of the shifting roles in Indigenous households – that is, that shifting roles were more than a reaction to high levels of male employment in the diamond mines – is made evident by the fact that all nine of the women who participated in interviews and initially self-identified as women whose partners worked in the diamond mines also worked in wage labour themselves on top of their primary roles in social reproduction and subsistence (this pattern was echoed in talking circles; however, because self-identification was not included in talking circle questions, it is not quantified here). Most of these women worked full-time jobs in Yellowknife or Behchokǫ̀; the rest were employed in temporary wage-labour positions. For example, Beth explained that she felt individually responsible for household social reproduction – including when her partner was home from the mines – even though she also worked full-time in a demanding job with a First Nations government. I asked her how she managed both her wage labour and the work of social reproduction:

My daughter goes to school so she's in kindergarten and my son is in day care all day. So I drop them off at day care, pick up my daughter at lunch hour. Then have lunch with her. Drop her back off at school. Pick her up at 3:30. Bring her to day care. And then pick them up at five. And when I get home I have to do to do all the laundry, all the cleaning, the supper. Bathe them …

[I ask her to mark out in a pie chart how much unpaid work her partner does and how much she does]

This would be the whole year [points to chart]. And this little chunk, that's one sixth of it, that's him. And this is me [points to 5/6ths of chart]. Because even when he's home, I still feel like I have to do everything. (Interview 107)

Accounting for northern Indigenous women's practices of social reproduction in relation to their own wage labour brings a specificity to their social location as departments of production shift as a result of the diamond mines. Kuokkanen (2011) reminds us that in many contemporary Indigenous communities in the Global North, women are often more involved in wage labour than men. Phoebe Nahanni (1992), writing of her home, notes that Dene's women's engagement with wage labour in the 1980s and 1990s, no less than men's, was accompanied by tension in relations of social reproduction, as well as sustained efforts to honour traditional Dene ways of being. In and around Yellowknife, the tensions between Indigenous women's central role in social reproduction and subsistence, and their extensive engagement in wage labour – both of which were discussed in almost all interviews – persist. This points to two insights. First, northern Indigenous women's engagement in capitalist production alongside social reproduction and subsistence should be approached in terms of the specific way in which Indigenous peoples in the region have been incorporated into Canadian capitalism – namely, that the Indigenous communities in and around Yellowknife have a limited history of engaging in social reproduction under the male-breadwinner/female-caregiver model. Indeed, prior to the diamond mines, while Indigenous men and women engaged with capitalist production in the mixed economy in different ways, there were no significant differentials in gendered rates of participation in wage labour or in remuneration. This is in contrast to the settler community that moved up to Yellowknife as part of the gold-mining boom and whose households, especially prior to government services moving north, were often structured through the male-breadwinner/female-caregiver model. Second, Indigenous women's responsibility for the household under the diamond-mining regime has been shaped by the neoliberal moment of capitalism – specifically, by the reprivatization of social reproduction (Bakker 2003).[8] This broad shift has interacted with local pressures so as to contribute to an intensification of household-level responsibilities, or the "making nuclear" of Indigenous women's social reproduction.

The diamond-mining regime's tendency to feminize and "make nuclear" social reproduction illustrates the contradictions of capitalist ideology and materiality, in practice, even as they develop in support of each other. The diamond-mining male breadwinner model that makes women responsible for caregiving does not match the history of Indigenous women's engagement in subsistence and capitalist production, nor does it correspond to the contemporaneous material need for women to continue this labour. As discussed in chapter 6, this contradiction is made manifest in the failure of the diamond-mining regime's efforts to recruit, train, and retain Indigenous women as mine workers. They are seen as a valuable source of labour, but because their social reproduction responsibilities are not taken into account (responsibilities that have been intensified by the same regime that wants to hire them), they are often unable to participate in that regime with consistency. The wage labour performed by Beth and other research participants points to the material fallacy of the "male breadwinner" model in the Yellowknife area. Many women provided stable wages for their family, whereas their partners' extractive wages had more of a tendency to ebb and flow. Yet it is undeniable that women take up the primary caregiving role, both out of necessity and out of social and interpersonal obligation. Rebecca Scott's (2007) work on dependent masculinity in coal-mining towns is useful here in understanding the contradictory dialectic between materiality and ideology. Scott takes up the local understanding of "work" in an Appalachian community with a long history of coal mining, wherein retail takes up approximately the same proportion of employment as the coal mines. Scott problematizes the naturalized understanding that (usually male) workers at the coal mines will make wages significantly higher than (usually female) retail workers, as well as the understanding that "real" work is the male breadwinners' work:

> A gendered understanding of work, embodied in the heterosexual white male breadwinner, gives shape to a specific configuration of masculinity that gains moral worth from family-wage employment. I am calling this formation of masculinity "dependent" in order to problematize dominant constructions of independence and dependence, public and private, that reproduce gender inequality. (486)

Scott (2007) offers insight into the ways in which the masculinized resource economy is a particularly intense manifestation of the male-breadwinner/female-caregiver model, even though women engage in wage labour at rates similar to those of men. But it is worth remembering here that, unlike the predominantly white community that had engaged in coal mining for generations in Scott's study, the diamond mines are something new, and the sustained push to integrate

Indigenous workers into the northern mining economy is also new. Many of the research participants did feel that their household caring responsibilities had both intensified and been normalized as a result of the diamond-mining regime, but they also talked about how they negotiated this responsibility through ongoing engagements with community subsistence (discussed in the next chapter) and maintenance of other waged and non-waged roles outside the home. For example, nine of the women interviewed and some of the participants in talking circles were engaged in paid work for First Nations organizations in Yellowknife and the surrounding communities. While those who had partners working in the mines discussed the difficulties of managing their paid work alongside their heavy care responsibilities, many of them also expressed great pride in how their organizations support their communities and protect northern Indigenous modes of life. Furthermore, as was not the case in the coal-mining town –Scott describes the masculinization of mining labour there as etched into the town's culture across generations – a number of women participants (like Doris) saw their non-participation in the diamond-mining regime not as an outside-enforced exclusion but as reflecting their own politics of refusal: they had chosen to prioritize waged and non-waged work that aligned with their responsibilities to their families, their communities, and their land.

Thus, just as Federici describes the gender violence of capital accumulation as a reaction to the power of women, rather than an amorphous phenomenon that denies their power, so too can contemporary violence targeting Indigenous women and girls be partly explained by their transformative and transgressive labours, as well as by the racist and sexist structures and ideologies that devalue these lives and labours. That is to say, violence against Indigenous women occurs not because of any deficit on their part, but rather the opposite: gender violence can be a racist and sexist reaction to power. A number of research participants discussed the gender violence that came with the restructuring of social reproduction, and they approached this subject in various ways. For example, in one talking circle, both women and men raised concerns about the Western versions of masculinity imported by the diamond mines – in particular, the pressures placed on young Indigenous men to enact a newly individualized responsibility for "providing" for the family and the potentially violent implications of this pressure, both in terms of men using physical violence against women partners and men engaging in self-harm. Angus:

> I worry about young men. There's a lot of conversations about other
> groups of people because young men come off as proud and as though

they're getting things out of this and having fun, but really, they're not. It hits someone at my age. I've had two suicide attempts and two full suicides of young men in the last seven or eight months. And it's almost like this hopelessness. And what's the long-term perspective? What's the great thing that I'm going to do? There's always this "be a man" kind of thing. It means that you're supposed to have an impact in some way, be important to somebody else. And we don't have these things. Maybe the language isn't there, things that people want to fight for. (Talking Circle One)

Angus was delineating the pressures placed on young Indigenous men to perform their role (as Indigenous people, as men, and as family members, partners, and members of their community) through a Western ideology of masculinity. While northern Indigenous men have always had specific roles in the community, the new – and newly masculinized – focus on capitalist production has repositioned "providing" for a family as individualized rather than relational labour. Angus followed up his first statement by discussing the potentially violent implications of men "failing" to enact the diamond-mining regime's construction of masculinity:

My biggest worry is that when men are failing, they damage other people. They can ruin people's lives really easily by being careless, by being greedy, selfish. And it's easy for men to fall into that because they have the power to have an impact on other people. (Talking Circle One)

Other participants in the talking circle echoed Angus's concern. Debbie:

I've also seen people with good jobs go out and get a nice paycheque and after a few paycheques, they buy a vehicle and then they get bigshot-ism. And then they lose their job. And they're worse than when they first started because their esteem is affected. And then they give up and don't want to do anything, so they take it out on the ones that they love most. Women are being brought to the hospital every day because of violence. And then that is transmitted to the younger ones. You see this violence, you know? (Talking Circle One)

Debbie and Angus both articulated the potential for violence in circumstances of "failure," when a man loses the dignity he once garnered by contributing to his community and household. At the same time, the potential for "failure" should be read through the larger settler political economy of the North and the uneven development that has contributed to the NWT's poverty and inequality. This is not simply on the doorstep of the diamond mines. As some interview research participants noted (Interviews 106, 302), employment in the diamond mines has also been a source of pride and stability for some men, contributing to their emotional *and* material well-being. What makes the FIFO diamond-mine regime distinct from other forms of wage labour in the region is that it has intensified and spatialized the notion of "providing" so as to reflect a hierarchical male-breadwinner/female-caregiver model. That model has the potential to inhibit traditional Indigenous social relations by invisibilizing land-based, care-based, and community-based modes of household provision while intensifying individualized pressures on both men and women to "succeed" in their roles. The specific instances of violence described by Angus and Debbie – suicide and gender violence – are embodied expressions of the racialized structural violence against Indigenous social relations, and their social reproduction.

In Canada, Indigenous women are three-and-a-half times more likely than other Canadian women to experience violence (Amnesty International 2009; CEDAW 2010; Sisters in Spirit 2010), and this figure is even higher in the North. The rate of violent crime against women – Indigenous and not – in the NWT is nine times the national average (Statistics Canada 2013). When it comes to extreme violence, the discrepancy between Indigenous and non-Indigenous experiences grows: in Canada, Indigenous women and girls are twelve times more likely to be murdered or missing than other women and sixteen times more likely than white women (MMIWG 2019). Too often, gender violence, sexual violence, and intimate partner violence are naturalized, approached as a tragic but inevitable and somewhat inexplicable aberration from the safe, non-violent norm. This naturalization is especially apparent in mainstream discourses of violence against Indigenous women in Canada. As Razack (2002) and Jiwani (2006) have argued, because of settler colonial processes of racialization, Canada's rural and urban Indigenous communities are often characterized as violent spaces, where violence is expected rather than explained. The 2019 National Inquiry into Missing and Murdered Indigenous Women and Girls has intervened in this dangerous act of concealment by naming violence against Indigenous women and girls and the 1,181 cases of "police-recorded incidents of Aboriginal female homicides and unresolved missing Aboriginal females" between 1980 and 2012 as genocide.

Naming this violence as genocide powerfully situates violence against Indigenous women as a central colonial strategy, not as something natural, not as an unavoidable "side-effect" (Anderson 2003). And just as women's role in socially reproducing a subsistence orientation helps explain the settler colonial tendency across time and place to target Indigenous women, in general, the specificities of the accumulation of new sites of extraction of surplus-value – in this case, in the form of diamond mining – intensify these violent tendencies, with the violence to the land and the violence to bodies a shared reflection of settler colonial accumulation. Symbolically, just as the Indigenous woman is deemed rape-able (Smith 2005, Kuokkanen 2008), so too is Indigenous land inherently explore-able and extract-able. Indeed, while there is often a purely geological understanding of sites of extraction – that is, an assumption that extractive projects emerge where there is a valuable resource to be extracted – a critical spatial and socio-geographic lens demonstrates that it is no coincidence that resource extraction consistently occurs in imagined frontier zones, so-called empty space, the spaces where sites of industrialization and urbanization have not encroached on Indigenous ways of being. Just as the idea of "wasting" the resources found in this imagined empty space is written as unthinkable, so, too, is it unthinkable to imagine uprooting the urban and industrial centres of Canada, no matter what might lie beneath. Thus, tearing up the concrete of the financial centres of the Canadian state is not considered; instead, the materiality of resource extraction is imposed on the symbolic periphery. Understood in this way, the long-standing centrality of resource extraction to northern economic development becomes a social phenomenon rather than a geological one.

The link between violence against Indigenous women and the violence of extraction is far more than merely symbolic. Actual sites of extraction are structurally violent on a number of levels. First, the physical act of extracting resources from the earth – which sometimes involves the use of poisonous chemicals – undermines the health and use-value of the land. Such practices harm any community close to an extractive site, and indeed, in the era of climate change, it would be hard to argue that there is anyone in the world entirely divorced from the environmental impacts of resource extraction. However, there is also a colonial specificity to the extractive violence done to the Dene, Métis, and Inuit land-based relations, or "grounded normativity" (Coulthard and Simpson 2016; Coulthard 2014). The symbolic, the material, and the corporeal are not divorced here. In an interview, Alica, a young Dene woman, spoke of a visceral feeling of emotional and spiritual sickness when she entered a mining camp to work. The two larger diamond mines in the NWT sit on her family's traditional trap line, where her grandparents lived out their days and where her mother was raised (Interview 119). Additionally, many informants spoke of the loss of country food (food sourced locally that is part of the traditional diet) as a result of the diamond mines, and their worries about the impacts this loss has had on emotional and physical health. For example, through the Orwellian-sounding "de-watering process," seven fish-bearing lakes were drained to make way for the first diamond mine (Bielawski 2003: 73–5), a practice that has been repeated for each mine. These were lakes teeming with life that Indigenous communities valued as sources of nourishment and well-being. At the same time, an even greater fear threaded through many interviews – the fear of cancer and its relationship to resource extraction.[9]

Alongside the links between physiological and emotional health and the land-based violence of extraction, there has been some (albeit limited) research on the links between the socio-economics of resource extraction and greater incidence of racialized gender violence. This research has emerged primarily from Latin America (Carrón-Prieto et al. 2007; Rodriguez 2012), and, more recently, Canada (Amnesty International 2016). Through its colonial imposition of gender relations that place masculinized activities devoted to/contributing to capital accumulation hierarchically over feminized and naturalized activities directed at social reproduction, the diamond-mining FIFO extractive regime enacts a structural violence. This structural violence persists dialectically through embodied experiences of racialized gender violence. A quantitative causative correlation (i.e., diamond mining leads to more incidents of violence against northern Indigenous women) does not necessarily follow from this relationship, in part because of the difficulties in reliably determining quantitative measures of violence against women, especially in the North, but more so because diamond mining is but one node of capitalist and Canadian state engagement with the mixed economy. These violences are an embodied articulation of the pressures weighing heavily upon labour oriented toward subsistence in the NWT mixed economy. I turn to this tension in the next chapter.

Conclusion

This chapter has examined the ways in which social reproduction performed by Indigenous women in and around Yellowknife has been restructured by the FIFO diamond-mining regime. That restructuring has occurred dialectically through the constrained choices of these women and the labours they perform. The diamond-mining regime demands an intensified orientation of social reproduction toward capital, facilitated by the historical processes of separating social reproduction from subsistence and the new spatial separation of social reproduction from capitalist production. By focusing on day-to-day labours, the preceding analysis sought to highlight the ways in which some Indigenous women's social reproduction has been "made nuclear" – that is, how the reorientation and restructuring of social reproduction toward the demands of the FIFO diamond-mining regime operate through a presumption of the (heterosexual) nuclear family, a presumption that has material consequences. The restructuring of the NWT mixed economy to meet the needs of the diamond mines is a violent structural process with embodied articulations, one example of the ways in which settler colonial processes of accumulation violently target Indigenous women and normalize that violence. However, while violence threads through the multiple scales of the social impact of extraction, social reproduction remains a key site of not only violent restructuring but also creative resistance in the regional mixed economy. In the final substantive chapter of this book, I analyse the ways in which community social relations oriented toward subsistence are being restructured by the diamond-mining regime, emphasizing corresponding processes of resistance.

8 Diamonds, Subsistence, and Resistance

I can't be part of this. I can't be part of something that is putting our animals and people and culture into the ground. For what? For a nice car? I was like, no. I swore off working in an office. I swore off working for a mine.

– Interview 116

The NWT skyline is streaked with flights moving mine workers from place to place through the work/home tempo of the FIFO regime. FIFO modes of extraction are a means of managing the tension between the fixity of resource extraction – in that, unlike in other industries, production (or at least the extractive portion of production) must be rooted in a particular place – and the neoliberal preference for spatial flexibility that can be shaped toward the temporal demands of capital (Rainnie et al. 2014). This tension, however, is just one layer of the contradictions between place and time that structure the NWT diamond-mining regime. The more fundamental contradiction is between the "grounded normativity" (Coulthard and Simpson 2016; Coulthard 2014) of the regional mixed economy and the intensified temporal capitalist imperative to produce surplus-value through, in the case of mining, the extraction of both labour and land itself.

Indigenous women's experiences of the (sometimes violent) tension between diamond mining and subsistence are the subject of this chapter. I argue that as a result of the diamond mines, there has been a shift in orientation away from land-based or subsistence relations, albeit a shift that is incomplete, contested, and shaped through resistance. While the diamond-mining regime has placed new pressures on their land and their lives, Dene, Métis, and Inuit women continue to deploy creative labours in order to protect and socially reproduce subsistence relations. In chapter 7, I analysed Indigenous women's labours primarily at the level of day-to-day household social reproduction; in this chapter, I approach these labours primarily at the site of community-level social reproduction, noting that there is not a firm boundary between household and community social reproduction and that the intersections between these levels are indicative of non-Western approaches to social reproduction in the region of study. Indeed, more than providing for daily social needs, Indigenous women's labour reproduces land-based relations in the mixed economy. This social reproduction is neither natural nor inevitable; rather, Indigenous women contribute to the reproduction of the mixed economy through constrained – often difficult or messy – choices and labours, shaped through the shifting structures of capital accumulation in the region as well as through relations of resistance.

The tempo of the FIFO diamond-mining regime has placed new pressures on day-to-day social reproduction; moreover, the temporariness of the regime has disrupted the intergenerational and community-level social reproduction of subsistence relations. In Section I, I take up the relationship between the diamond mines and subsistence, exploring the diamond-mining regime's approach to subsistence activities and contrasting this with research participants' analysis of the relationship between FIFO and subsistence. *Contra* the collective histories, relationships, and ethics – the "grounded normativity" – through which subsistence activities are practised in the northern mixed economy, the extractive regime takes an individualized, dehistoricized approach to subsistence, one that enables the industry to claim that the FIFO work tempo can support and even facilitate subsistence (GNWT 2016). While certainly, some individuals can and do engage in land-based activities in their weeks away from the mines, I suggest that the broader impact of the FIFO tempo in particular, and the demands of mine employment in general, is rather that these things interrupt and disrupt the intergenerational transmission of the knowledge and social relations required for subsistence. In Section II, I offer an alternative approach to examining the rupture enacted by the diamond mines at the site of community-level labours of social reproduction and subsistence. Drawing on ethnographic analysis of contemporary subsistence production in the NWT, I bring discussions of the regional characteristics of subsistence in conversation with Maria Mies and Veronika Bennholdt-Thomsen's (1999) articulation of a concept they refer to as the "subsistence perspective." That concept provides a theoretical tool for elevating labours performed in the context of global capitalism, labours that, while potentially participating in the production of surplus-value, maintain an orientation toward land-based relations. This approach to subsistence – as an orientation rather than a fixed typology – offers analytical space for the relational nature of the negotiation of rupture, as articulated in Tłı̨chǫ cosmology. Analysing subsistence as a relational concept, in Section III I explore the strain the extractive regime places on Indigenous women's subsistence labours by examining the impact of the diamond mines on inter-household sharing, place-based education, and the health and integrity of the land and the body. In so doing, I show that the temporary rupture that is the FIFO diamond-mining regime has presented new challenges to the intergenerational reproduction of a subsistence orientation in the mixed economy. These challenges are negotiated by research participants through the tight connection between diamond mining and structural and embodied violence, a connection that is wound up in modern histories of resource extraction, subsistence, and resistance, as well as forward-looking concerns for future generations. Extending analysis to the community level links Indigenous women's daily labours to the social reproduction of the mixed economy, making visible the many ways in which those women continue to resist the totalizing impulses of the Canadian state and capital. In examining the diverse ways in which subsistence relations are reproduced, it is notable that while at the theoretical level, subsistence and capitalist production are antithetical, at the level of social formation – that is, in the grounded ways in which people perform their day-to-day labours in the region – the relationship between capitalist production and subsistence is tightly interwoven, while also tenuous and contradictory. As the research participants share in the narratives that inform this chapter, northern Indigenous peoples have long engaged in wage labour to support subsistence activities (Usher et al. 2003, Abele 2009a, Harnum et al. 2014), and non-Indigenous and Indigenous peoples have worked together to build local socio-economies that complicate and extend beyond the assumed settler/Indigenous binary (Abele 2015; Southcott, ed., 2015), exemplifying the Dene relational approach to development (Zoe in Gibson 2008) that extends to both different peoples and the land on which they gather (Coulthard 2014).

I. The Diamond-Mining Regime and Subsistence

As Usher and colleagues note (2003), subsistence production in the traditional and mixed economy has a history of being underresearched in settler socio-economic studies,[1] not least because, up until the 1970s, social policy was driven by the assumption that subsistence production would simply disappear as a consequence of "development" (176). However, after decades of organizing and advocacy, today the GNWT recognizes subsistence production as part of the NWT mixed economy. The GNWT collects community data on "traditional activities" and funds in-depth investigations into certain traditional activities in specific regions (GNWT 2015a); that these data are being collected speaks to the territorial government's recognition – at least – of subsistence as a contemporary economic activity. This hard-won territorial attention to subsistence activities interacts with the diamond mines' interest in, and approach to, Indigenous engagement to produce sustained, albeit limited, monitoring of the diamond-mining regime's impact on subsistence. Specifically, under the SEAs, the GNWT, partnered with the diamond-mining companies, is drawing upon territorial data tracking subsistence, as well as other socio-economic and public health indicators, to measure the impact of the diamond mines on Indigenous communities and their traditional activities and well-being. This analysis is published in annual *Communities and Diamonds* reports.

Reflecting the diamond-mining regime's marketized approach to Indigenous/settler relations, SEAs approach subsistence along individualistic lines. Subsistence is conceptualized as a cultural recreational activity rather than a relational and ontological practice. It is no surprise that subsistence is characterized in this way: while subsistence imagined as recreation is complementary to the FIFO tempo insofar as individual workers can hunt or fish on their time off, for example, subsistence as place-based and relational is antithetical to both the FIFO tempo and the temporariness of the regime itself. In state and industry analysis, that tension is lost. What I mean here is that, materially, the diamond mines degrade the land on which they rely, thereby endangering present and future subsistence practices. Simultaneously, by taking workers out of their communities, the diamond mines make the ongoing knowledge transmission necessary for subsistence more difficult. The displacement of earth and animals – in particular, the interruption of caribou herds by mining roads that cut across their migratory paths – as a result of the diamond mines degrades the already taxed northern landscape, thereby threatening the persistence of subsistence on this fragile land. When community members are away at the mining camps, they are unable to participate in communal practices like hunting and trapping, preserving meats, tanning hides, and picking medicines. At the same time that land and animals are threatened, so too is the knowledge transmission required to reproduce subsistence practices. Thus, rather than complementing subsistence production, the FIFO diamond-mining regime has initiated a rupture in local social relations, and this has the potential to threaten subsistence relations in the mixed economy.

How, then, do state and industry approach the relationship between the diamond mines and subsistence? The GNWT measures subsistence in terms of reported engagement in specific "traditional activities." According to the GNWT, in 2013, 44.7 per cent of the territorial population hunted and fished; 6.1 per cent of the population trapped; 23.3 per cent of the population produced arts and crafts; and 26.3 per cent of Indigenous households relied on country food for at least half their consumption (GNWT 2015a).[2] It is notable that these figures represent the activities of both the Indigenous and the non-Indigenous territorial population; it follows that the larger percentage of Indigenous people in the territory engaging in subsistence production is not captured. Though it is beyond the scope of this overview, it would be a revealing exercise to further reflect on the non-Indigenous people engaging in subsistence activities: practices that speak to the complex interdevelopments of the mixed economy.[3] These data are also collected at a community level, though, as noted in chapter 2, N'dilo is counted as part of Yellowknife, and, correspondingly, its distinct relationship to subsistence activities is not captured. In 2013, in Yellowknife, the percentage of the population that engaged in arts and crafts was 19.6; in Dettah, it was 30.9, and in Behchokǫ̀, 25.4. Unsurprisingly, Yellowknife, as the space with the highest settler population, had the lowest percentage of residents engaged in traditional arts and crafts, with most communities' participation rate ranging between 25 and 40 per cent and the highest participation rate being 54.4 per cent (in Jean Marie River, a small community of less than 100 people). In both Dettah and Yellowknife, 37.1 per cent, of the population hunted or fished in 2013; in Behchokǫ̀, 40.5 per cent. In Yellowknife, only 2.1 per cent of the population trapped; in Dettah and Behchokǫ̀, 15.5 and 10.7 per cent respectively (GNWT 2015a). In thinking through these data, recall both the ways in which people move between urban centres and their home communities and the ways in which urban centres depend on the subsistence production of rural communities (Abele 2009a). These dynamics cannot be captured by the numbers.

Before the mines become operational, as part of their SEAs, companies are asked by the GNWT to "*anticipate* the impact their diamond mine may have 'on the people and communities of the NWT'" (GNWT 2013a: 1) by commenting on a series of indicators developed by the territorial government. While subsistence is not an indicator unto itself, "cultural well-being" and the "traditional economy" are two of the areas on which the diamond companies are asked to comment; both speak to subsistence. All three diamond-mining companies noted the potential negative impacts of the diamond mines on subsistence: BHP predicted that the impact would be "negative, but small." Diavik wrote that "wage-based activities may erode ... Dene, Métis and Inuit Culture" (GNWT 2013a). Notably, De Beers, writing about the potential impacts of the Snap Lake diamond mine, claimed that

> the limited amount of time in the community may limit individuals' ability to pursue Aboriginal traditional activities, which impacts on individuals' lifestyle and the maintenance of a cultural identity ... The family as a whole will also be affected by the limited time available to engage in traditional activities with all family members present. This may complicate efforts to maintain cultural traditions and identity. (GNWT 2013a: B-5)

De Beers – whose assessments of the impacts of Snap Lake and Gatcho Kué are by far the most recent – is articulating here an awareness of the interpersonal impacts that an individual's employment may have on subsistence activities. It is notable, too, that the corporation describes potential impacts on the household rather than the individual worker.

The predicted social impacts of the diamond mines are monitored through GNWT community statistics, like those shared above, and reported through the annual *Communities and Diamonds* report, compiled by the GNWT with the participation of the diamond companies. Up until 2015, the *Communities and Diamonds* report tracked trapping, hunting, and fishing and noted the percentage of households reporting that half or more of the meat or fish they consume is harvested in the NWT. These data are collected through the GNWT and disaggregated between the towns and communities of the NWT. While the working government definition of the traditional economy does include harvesting and "making crafts by using raw materials from the land" (GNWT 2014a), these activities are not tracked or reported on in the *Communities and Diamonds* reports. This gendered omission has contributed to the tendency toward obscuring feminized subsistence labour (including harvesting, craftwork, sewing, and preparing land materials for household use and consumption), which serves to perpetuate a focus on individualized, masculinized activities (namely hunting, trapping, and fishing).

The following graphs, based on NWT data, report trapping, hunting, and fishing activities and the consumption of locally harvested food from the period preceding the first operational diamond mine (which opened in 1998) to the period of the most recent available data at the time.

The data show no significant changes after the first diamond mine opened in 1998 or throughout the growth of the industry in the decade

Figure 5. Percentage of persons 15 and over who trapped in a given year, by region; Northwest Territories, 1988–2018. (2021 NWT Bureau of Statistics).

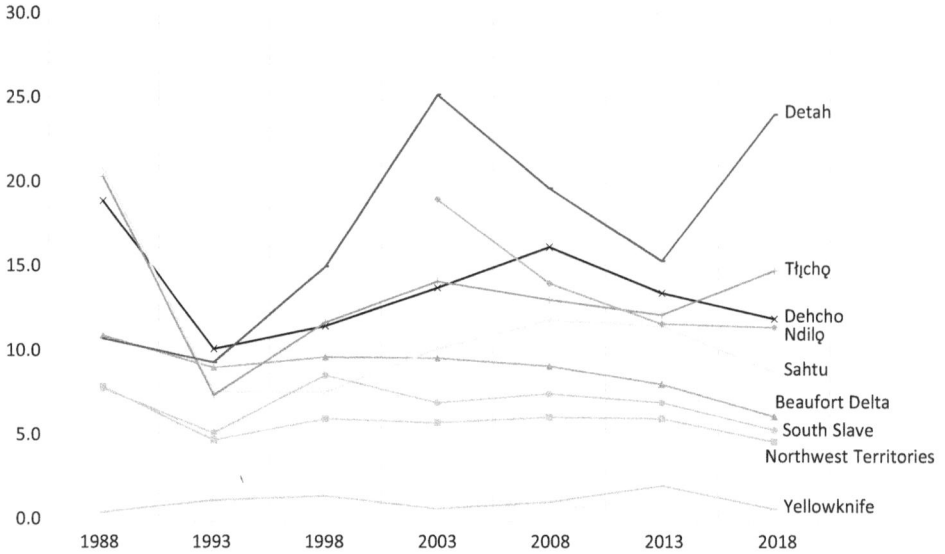

that followed. Yellowknife, again, consistently falls at the bottom in terms of tracked subsistence activities. Based on the data illustrated here, the *Communities and Diamonds* 2014 report concludes that hunting, fishing, and eating country food "does not seem to be influenced by mining" (GNWT 2014a: 27). The report does note that trapping has declined since 1998 – the year that the first diamond mine opened: "mine employment does seem to be affecting this change, with fewer young people willing to pursue employment in the traditional activities" (GNWT 2013a: 28). However, the possibility that diamond mining is interrupting intergenerational transmission of subsistence knowledge and skills is entertained only momentarily: the sentence that follows is this: "It is possible there is a link between jobs at the mines and having money to get out on the land during time off work (Behchokò ... [which, in exception to the trend, has had higher trapping rates since the diamond mines opened] ... is an example of this)" (GNWT 2013a: 28). This suggestion – that work in the diamond mines leads to greater access to hunting, trapping, and fishing – was echoed in interviews with government workers and diamond mine management.

Figure 6. Percentage of persons 15 and over who hunted or fished in the year, by community, Northwest Territories, 1998–2018 (2021 NWT Bureau of Statistics).

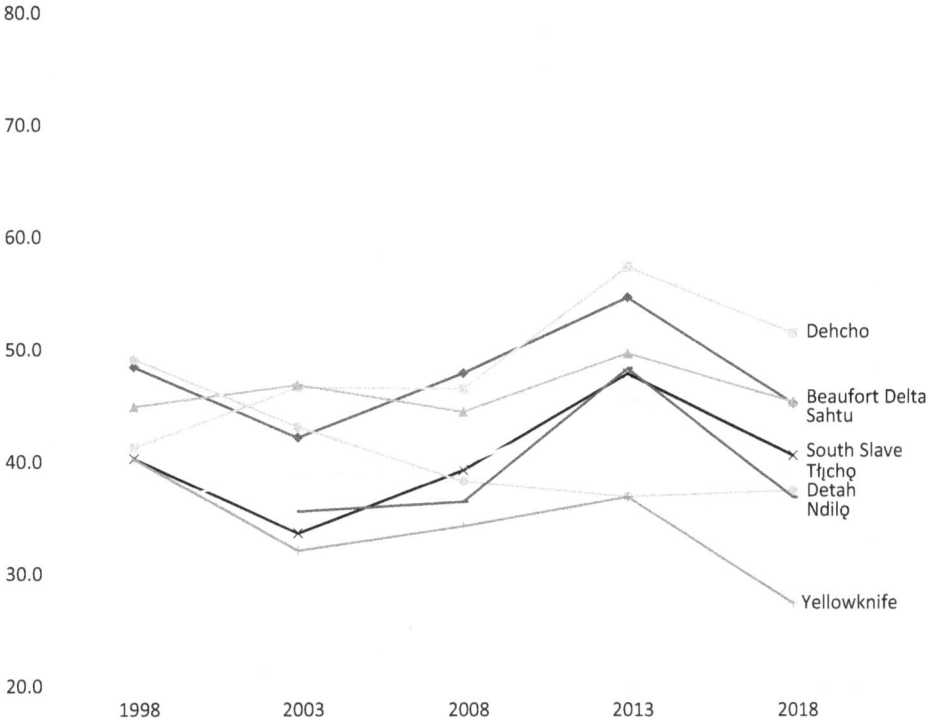

Because they rely on quantitative analysis without supporting analysis from the communities themselves, the *Communities and Diamonds* reports seem to oscillate between making diamond mines the sole causative factor for all observed social shifts and denying responsibility for any troubling social trends (such as rising suicide rates) by selectively noting the limitations of a single-cause approach (GNWT 2014a). Given the relational quality and the grounded histories, practices, and ontologies that shape subsistence – and the gendered variation in experiences with and responsibilities for subsistence activities – the quantitative data alone tell us little about the relationship between the diamond mines, subsistence production, and potential shifts in orientation in departments of production.[4]

Figure 7. Percentage of households where 75% or more (most or all) of the meat or fish eaten in the household, by community, Northwest Territories, 1998–2018 (2021 NWT Bureau of Statistics).

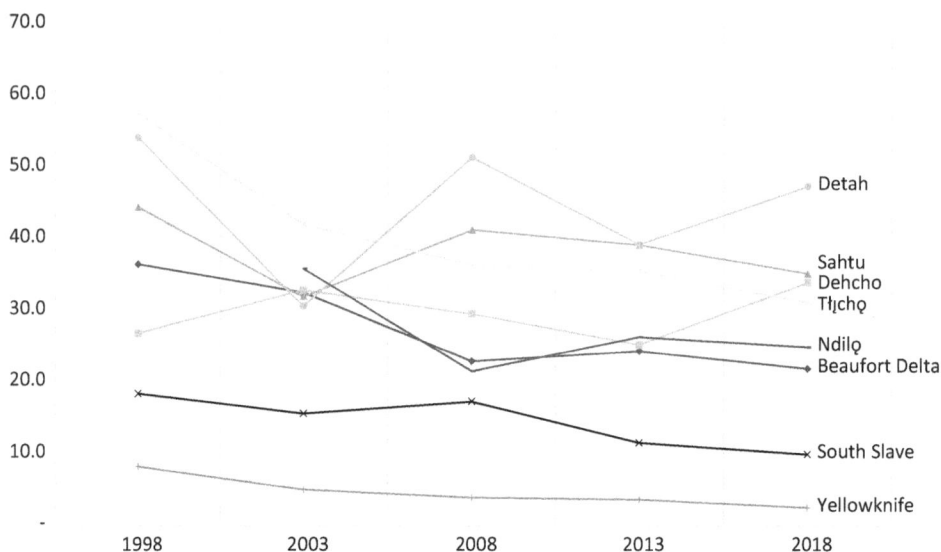

Instead, the *Communities and Diamonds* reports are a telling articulation of the general approach to subsistence taken up by the diamond-mining regime, an approach that perpetuates and intensifies masculinist, individualized conceptions of subsistence production. Under the diamond-mining regime, the general tendency to masculinize and individualize subsistence has taken a particular form: the understanding of subsistence as recreation. Specifically, subsistence production is translated as individualized acts of recreation undertaken by male mine workers in their "time off." In this conceptualization, subsistence production – referred to as "traditional activities" in industry documents and understood primarily as hunting and fishing, and sometimes trapping – is removed from its relational, community quality (Coulthard 2010), as well as from its role in the intergenerational reproduction of local Métis, Dene, and Inuit social relations. In this formulation, the FIFO structure allows diamond workers to hunt and fish during their time off, and relatively high salaries give workers more opportunity to buy gear that can help them in these pursuits (GNWT 2014a: 27). Thus, subsistence as an activity is racialized as a symbolic marker of "being Indigenous" and masculinized through a focus on activities traditionally undertaken by men at the expense of activities traditionally undertaken by women, or activities – such as meat preparation – that illustrate the gendered and inter-household interdependence of subsistence activities. So at the same time, subsistence is materially divorced from its role in socially reproducing Indigenous social relations. By approaching "traditional activities" in this way, separate from attachments to places and people, the extractive regime is able to assert that the FIFO diamond-mining regime supports local Indigenous practices.

Certainly, the general point that money garnered through wage labour can be used to support subsistence activities reflects a long history in the North of Indigenous people engaging in wage labour in this way (as noted by Usher et al. 2003; Abele 2006; Harnum et al. 2014; see below). However, the *particular* suggestion that the FIFO tempo of the diamond-mining regime supports subsistence production is based on an individualized interpretation of subsistence that denies the ways in which interdependent social relations in Indigenous communities are being undermined by the FIFO diamond-mining regime. Much like the diamond mines' approach to Indigenous/settler relations, this dehistoricized approach to subsistence enables an illusion of an "Indigenous-friendly" extractive regime, obscuring the irony of an extractive regime – built upon the temporary exploitation of a specific land mass – positing itself as a support to the subsistence activities rooted in that same land. In this way, "Indigeneity" as a racialized category is subtly reconfigured to support the extractive project at hand. Indeed, for the diamond-mining companies, if "being Aboriginal" simply means engaging in acts deployed as markers of Aboriginal-ness (as Gibson [2008] critiques), then the diamond mines are exempt from criticism for disrupting Indigenous socio-economies through FIFO labour and extraction. And indeed, in interviews, industry and government supporters of the diamond mines further pressed this point by comparing FIFO labour with the nomadic seasonal labour patterns of northern Indigenous peoples and, more specifically, traditional seasonal hunts that would see men separated from their families for long stretches of time. As Emma, employed in diamond mine management, said, "So we can get too obsessed with two and two [two weeks on/two weeks off]. There was this whole other world of seasonal work, and just, quite frankly, trapping. Guys would be gone for months. That was just part of the seasonal rounds" (Interview 402). For her, then, given that Indigenous peoples have a history of travel and separation from kin (largely men leaving for long periods of time), FIFO does not introduce a new dynamic.

While a number of research participants did point to the possibility of mine wages paying for the materials needed for land-based activities, for many of the Indigenous women who participated in interviews and talking circles, the suggestion that FIFO work could be equated with seasonal activities rang false. They saw interdependent, seasonal, communally driven hunting expeditions as distinct from individualistic hunting trips, for example, scheduled around FIFO work. These women noted that FIFO schedules were more likely to disrupt or disallow community-organized seasonal hunting, fishing, and gathering or preparation of goods. For example, when discussing the impact of the diamond mines on subsistence production, Alica said:

> P: I remember in a meeting with one of the mines once, I forget which one, one of the representatives from the mine made some comment, said, "Actually, it's a good thing – the two weeks in, two weeks out – because it represents the traditional way of life where the man goes away to go hunting." I was like, are you serious? It's not how it works.
> R: That's funny. I've also heard that.
> P: He must have just thrown it out there. It doesn't work like that. I have never heard an Aboriginal person say that, "Oh, this is great. This is totally how it was when my parents grew up." (Interview 119)

For Alica, the notion that FIFO work was structurally similar to traditional hunts denied the meaning, the relationships, and the *quality* of subsistence production. Traditional hunts and other subsistence activities require that people build their labour and their travel around the needs of the community and the seasonal imperatives of the land. FIFO, by contrast, imposes a rigidity determined by the schedules of the extractive regime. Similarly, Annie, a Dene woman and community worker, described the FIFO schedule as disruptive of subsistence production rather than supportive of it: "I know that people still go out hunting and that. But as I said, there's an absence of a mother or a father. Or the woman has to wait for two or three weeks. If the father's going to go out hunting, they have to wait" (Interview 207).

Angela, a younger Dene woman with many male family members working for the diamond mines, also discussed the ways in which working for the mine detaches people from the tempo and relations of subsistence production. I asked her if her family members who work for the mine (including cousins who work for the mine, all of whom were male) engaged in subsistence production:

I can verify from my family, not so much. Like my brother really enjoys hunting and stuff, but he doesn't go as much as he would want. And he doesn't have children or a girlfriend at this time ... And same with my cousins. I don't see them going out on the land as much. Like, one cousin, he was in-between jobs, and when he was in-between, it was maybe a six-month period, and by the end of that six-month period, he was finally at a place where he was being proactive and going out on the land. But I think it just took getting him to a place where he was bored, he had no money to spend. It was survival. It wasn't cultural. (Interview 303)

Angela's last point speaks to the importance of looking to the materiality and the relations through which subsistence production is performed in order to understand its shifts over time. In characterizing her one cousin's subsistence practices as survival, particularly in a context of a highly unequal mixed economy, Angela's anecdote is a reminder not to romanticize labour perceived as "traditional," but rather to understand the tensions between subsistence and FIFO wage labour in the context of an uneven settler economy.

These narratives intervene in the assertion by the diamond-mining regime that the diamond mines do not adversely impact subsistence, but rather support it. However, while the gaps in the suggestion that work in the diamond mines supports subsistence are clear, the inverse suggestion – that engagement with resource extraction is singularly harmful to subsistence production – does not, in and of itself, paint a full picture of the relationships among departments of production. Abele (2006) reminds us that in the northern mixed economy, wages from capitalist production have become necessary in order to purchase items required for subsistence practices, such as snowmobiles and guns. And in a report on the relationship between extractive activities and subsistence, Dene residents of the Sahtu region explained that the households that thrive in subsistence production are often the same households that do well in the wage economy (Harnum et al. 2014). Harnum and colleagues contend that the sharing of resources within and among households enables families to benefit from "both the wage economy (used to support subsistence activities) and subsistence activities" (28). Usher and colleagues (2003) make the same argument: "The successful harvesting household is often also the successful wage-earning household, as this cash income is used for purchasing harvesting equipment, and especially fast means of transport" (178). Harnum and colleagues' report is speaking to the dynamics of the Sahtu region, a more rural part of the NWT, and Usher and colleagues are speaking about the Circumpolar North in general, so it is worth calling attention to the distinct political economy of the Yellowknife region. Yellowknife and to some extent the nearby towns (Behchokǫ̀, Dettah, Ndilo) host a relationship between subsistence production and wage labour different from that of more rural parts of the territory – namely, capitalist production has a stronger presence around Yellowknife than in smaller communities. In 2016, for example, the average rate of capitalist employment in the NWT was 69.2 per cent, but it was 77 per cent in the Yellowknife area (including Dettah and N'dilo). For comparison, the rate was 59.3 per cent in the Sahtu region and 40.6 per cent in the Tłı̨chǫ region (which includes Behchokǫ̀). The Tłı̨chǫ employment rate points to a disparity in employment rates in and around Yellowknife – the city of Yellowknife's employment rate is 77.6 per cent, compared to 44.4 per cent in Dettah and 36.4 per cent in Behchokǫ̀ (Ndilo's employment rate is not disaggregated from that of Yellowknife) (GNWT 2016).

Abele (2006), Usher and colleagues (2003), and Harnum and colleagues (2014) rightly point to the ways in which wages garnered from capitalist production can support subsistence production, and also to the increased social need for capitalist wages to support subsistence in the context of the mixed economy (i.e., the diamond mines reflect a history of capitalist penetration that has eroded what had once been entirely a subsistence economy). However, the studies by Harnum and Usher their colleagues take the household as the unit of analysis, and this obscures the ways in which the triadic relationship between capitalist production, subsistence production, and social reproduction is gendered. I explore that gendering in the rest of this chapter, first through a discussion of the characteristics of production and social reproduction oriented toward subsistence in the mixed economy, and then through an analysis of subsistence in relation to the diamond-mining regime.

II. A Subsistence Orientation

In analysing contemporary subsistence practices, northern scholars have emphasized the crucial role of sharing in production and distribution within and across households (Coulthard 2014; Harnum et al. 2014; Abele 2006; Kulchyski 2005; Usher et al. 2003). There is a tendency to "nuclearize" Indigenous relations of social reproduction through Canadian state and extractive interventions. However, the push toward nuclear-level social reproduction exists in the context of, and in tension with, strong histories and practices of community-level sharing (Dombrowski et al. 2013). Harnum and colleagues (2014) explore the ways in which the goods gathered and made by subsistence labour are shared, not just through direct networks, but among friends, extended kin and community networks, and people in need: "In a survey conducted by the Dene Nation [in 2010], more than 60% of people in Dene communities in the NWT said they often received *dene bere* (country foods) from others, while 37% indicated that this occurred occasionally" (20).[5] Participants in Harnum and colleagues' research project described subsistence production as an activity that enabled "more family and community cohesion and a greater sense of pride and identity" (52). Indeed, subsistence production is undertaken in interdependent ways across households, and this contributes to the well-being of the whole group. Indigenous activists and scholars have long pointed to the non-nuclear structure of care in Indigenous households (see, for example, Nahanni 1992; Anderson and Lawrence 2003). Usher and colleagues (2003) describe this as "supra-household interaction" and note that it is determined largely by kinship networks. They write that these connections are, "celebrated, consolidated, reinforced and reproduced by sharing, feasting, ritual observance, and associated ethical norms. There is much incentive to maintain the system, little to disrupt it" (179). Similarly, participants in this research often emphasized inter-household connections – strong kin and community networks within which social reproduction and subsistence are undertaken interdependently across nuclear family units. Alica, describing her own extended family, said that "I grew up around little kids, helping take care of the family. But that's pretty common. And that's a big part of Aboriginal culture, too, family raising the family" (Interview 119).

When analysis focuses on the household level, the labour category ("social reproduction") is distinguishable from production (especially in relation to capitalist production, but also in relation to subsistence). Conversely, labour performed across households challenges the separation between subsistence production and social reproduction. For example, when Indigenous women prepare caribou meat and share it among their households, this is both subsistence and social reproduction (and thus is referred to in this study as social reproduction oriented toward subsistence). Recall that the separation between social reproduction and subsistence is a settler colonial construct that only partly shapes the social relations of the mixed economy. However, while these inter-household networks are spaces of decolonizing relations, it does not follow that they are inherently nurturing, nor does it follow that they are free of their own unequal power relations. Indeed, just as feminist and anti-racist analyses have challenged depoliticized relations within the household, so, too, must inter-household and kin relations be politicized rather than romanticized. These relations are a key site of struggle in the maintenance of a subsistence orientation within the mixed economy.

Similarly, subsistence relations are enacted and strengthened through intergenerational learning. Subsistence production is a form of knowledge- and skills-sharing rooted in the land and linking generations; it is also a primary form of community-level intergenerational social reproduction in the NWT (Nahanni 1992). This is a good illustration of the complementary and at times indistinguishable nature of subsistence and social reproduction. Subsistence as a means of producing or acquiring communal needs is often at the same time a means of intergenerational education and of reproducing the ethics integral to "grounded normativity" (Coulthard and Simpson 2016; Coulthard 2014) and, thus, to social reproduction. Kuokkanen (2011), drawing on the work of Eugene Hunn (1999), writes that "there is a crucial link between subsistence and Indigenous knowledge" (220). Similarly, in speaking about "*Naats enelu he asii yats'ihtsi* (arts and crafts)," the participants in *The Best of Both Worlds* noted that engaging in subsistence activities provided

> much-needed *deneghagot'a* (opportunity) for the intergenerational transmission of not only skills for producing articles, but also *deneghaghot'a* to share *dene nawere* (traditional knowledge) with young *ts'eku ke* (women) in particular. For example, a local "sewing circle" involves not only *dene ghaonete* (teaching) skills for *naats'enelu* (sewing), but also creates a venue for passing on *dene naowere* about producing raw materials such as sinew and hides, traditional *ewa taadenakwi* (hide-tanning) methods, child rearing, caring for the sick and *ohda ke* (elderly) roles of *deneyu ke* (men) and *ts'eku ke* and related custom, kinship, *denewa naridii* (traditional medicines) and so on. (Harnum et al. 2014: 58)

Using as an example a sewing circle (an activity that is traditionally undertaken by women), Harnum and colleagues demonstrate the important roles women take up in socially reproducing the knowledge, values, skills, and relationships needed for a subsistence orientation. Nahanni (1992) draws upon her experiences on the land with Dene families to demonstrate the tight links between subsistence and intergenerational education. She followed family members who work with their children on the land to teach them the skills they need to reproduce a subsistence orientation. Notably, in discussing Dene women's teaching role, she links their subsistence activities with their educative and caring responsibilities, responsibilities that FPE associates with social reproduction. She demonstrates that "'nurturing' and 'providing' are customarily attached to Dene women's approaches to 'learning' and 'teaching'" (4).

Research participants in this project echoed similar themes. For example, Liza, an Indigenous woman and a community leader, described her subsistence work as intergenerational social reproduction:

> We still do those kinds of things. And I have done a couple of presentations on medicine. And we've gone to pick medicine. We'll probably go this fall to pick medicine. And I tell the ladies, when you go to pick medicine, pick it and we come back and prepare and preserve it, now it's your responsibility to pass it on. Because nobody else will do it for you now. I'm not going to be here. Pass it on to your kids and your family. Make sure they know that there's good medicine out there. You know, that a lot of people don't know about. (Interview 206)

For Liza, the subsistence work of picking and preserving medicines was of a piece with the transmission of the knowledge it required. Similarly, Tory, a young Indigenous woman, identified her learnings in the bush from uncles and aunties as an adolescent as formative educative experiences. She compared her own experience of education in her adolescence with those of her non-Indigenous classmates:

> I was working in the bush most of my summers. And that was another one, working in the bush for my summers, because when I got back to school all these kids would tell me what they did in the city. Going to Edmonton, doing all these other things. And I was like, "What are you talking about?" I'd get really embarrassed because I was like, "I just made a bunch of dry meat all summer in the bush." You know, nothing that crazy. (Interview 116)

Tory joked about the disparity between her summer vacation in the bush and the shopping trips of her peers, and noted feelings of difference and discomfort. But she also recounted with pride the stories and lessons shared by her aunties and uncles. Tory saw her work in the bush (both as an adolescent and as a young adult) as crucial to developing relationships, knowledge, and skills that she viewed as essential to her being. She also recounted early experiences of racist violence, such as being bullied and beaten up for being Indigenous in the schoolyard as a child. Her experiences of racism threaded through her description of adolescence, and she named subsistence education as an important source of healing from racism, of both learning and pride.

Tory grew up in Yellowknife, and her work and education history offers an example of a life narrative that has moved fluidly between subsistence and capitalist production. Similarly, Liza's work in community education and service provision has moved between that which has been "professionalized" (paid, supported by state institutions) and that which has emerged through community relations. For both women, what remained consistent in the multiple labours they described was an orientation toward subsistence. The concept of a "subsistence perspective" (Mies and Bennholdt-Thomsen 1999) offers a means of honouring this orientation and assessing shifts in orientation and power in departments of production, rather than imposing a binary (*either* subsistence *or* capitalist production). Such an orientation can account for the dynamic interplay between departments of production in the North, especially through the processes of rupture – a person might take up wage labour to pay for the tools he or she needs for subsistence practices, or an urban household might fill the gaps in paycheques from a non-steady job through the country food provided by their family living in a rural community, for example – while at the same time acknowledging and elevating unequal racialized and gendered power relations and the violence that can accompany shifts in the power and place occupied by capitalist production within the mixed economy.

Mies and Bennholdt-Thomsen (1999) developed the "subsistence perspective" as a feminist analytical tool aimed at naming and gendering contemporary production oriented toward meeting social needs rather than the accumulation of capital. They wished to reclaim the concept of subsistence from the faulty premise that subsistence is either "merely survival" or, as a result of global capitalism, no longer exists. Echoing analyses of the role of subsistence in the North, they write that "subsistence not only means hard labour and living at the margins of existence but also joy in life, happiness and abundance" (5). They suggest that the fact of capitalist production is not necessarily followed by a capitalist orientation and that a "subsistence perspective" (i.e., labour oriented toward the social reproduction and well-being of the collective rather than the profit of the one) is a powerful site of anti-capitalist, feminist resistance. A subsistence perspective "insists on the priority of use-value production" (58) and prioritizes the "creation and maintenance of life on this planet" over "the accumulation of dead money" (7). And indeed, in the northern mixed economy, where capitalist and subsistence production coexist, and social reproduction is undertaken relationally with both forms of production, the question is not whether a particular form of labour is performed but under what conditions, through what systems of power and meaning, and to what end.

Northern Indigenous peoples have long engaged with wage labour through an orientation toward subsistence. For example, Blondin tracks his family's history of mediating the "two worlds" of traditional

Dene activities and the new settler economy (1990). Harnum and colleagues (2014) write that "many people seek *eghalaeda* (jobs) as a means of providing the cash needed to equip themselves and their families for traditional activities such as *nats'eze* (hunting), *ehdzo ats'eh* (trapping), *dats'e* (fisheries), *leetehele* (gathering) and obtaining materials for the production of traditional *naats enelu he asii yatsihitsi* (arts and crafts)" (64).[6] Usher and colleagues (2003) come to a similar conclusion in their discussion of northern mixed economies, arguing that individuals, households, and communities do not choose between subsistence and capitalist production but rather negotiate their location within an economy that encompasses both: "In subsistence-based systems, the ends of economic activity tend to be inseparable from the social system, and are more likely to be the maintenance of the system of social relations rather than accumulation at the level of enterprise" (179). Around Yellowknife, where capitalist production has a long history as well as a stronger presence than among the Sahtu and in many other regions of the NWT, a subsistence orientation is more the *subject* of questioning than the conclusion; however, these insights speak to the specificity of the orientation and meaning of labour in the mixed economy. A crucial message underlying the Tłı̨chǫ concept of rupture is that the imposition of new social relations is never total; rather, it is in constant negotiation and relation with the modes of life established throughout the long history of northern Indigenous peoples.

Indigenous women in the NWT have been uniquely targeted in the steering of social reproduction away from subsistence toward capitalist production. The gendered shifts in the subsistence orientation in the northern mixed economy exemplify Mies and Bennholdt-Thomsen's (1999) argument that modern threats to subsistence are threats to women, and that this is because of women's central role in subsistence production.[7] They link the de-economization (which I take to mean the obfuscation of subsistence labour, the use-values it contributes to society, and the ways in which capitalist production relies on its exploitation) of subsistence labour with that of women's labour: "The de-economisation of female labour and the de-economisation of subsistence are one and the same process" (115). By emphasizing the feminization of non-capitalist labour mobilized for the pursuit of capitalist exploitation, Mies and Bennholdt-Thomsen offer a compelling approach to the restructuring of northern Indigenous women's non-capitalist labour. They note that, while the exploitation of wage labour is facilitated structurally through the capitalist mode of production,[8] the exploitation of non-capitalist labour (what they call subsistence, which in the terminology of this book refers to both social reproduction and subsistence) requires violence and coercion.

This insight connects to Mies's earlier work on the gendered violence of primitive accumulation (1986). When gendered violence is linked to exploitation and oppression at the site of both social reproduction *and* subsistence labour, it becomes clear that embodied violence for the pursuit of capital manifests itself differently – and, I argue, more intensely – upon the bodies of women engaged in subsistence-oriented labour. This speaks to a continuity rooted in place and across modern time in the ways in which Indigenous women in the NWT are particularly targeted by the violence associated with capital accumulation of new spaces. As Smith (2005), Kuokkanen (2011), and Simpson (2016) argue, Indigenous women's bodies are often materially and symbolically linked to the land, and have been "rendered less valuable because of what they are taken to represent ... alternatives to heteropatriarchical and Victorian rules of descent" (Simpson 2016). Through settler colonial processes of racialization, structural and embodied violence against Indigenous women in Canada has been made permissible. This is a violent racism that crosses the rural and the urban as well as the diverse regional political economies of Canada. Thus, while Indigenous women's role in socially reproducing a subsistence orientation helps explain the settler colonial tendency across time and place to target Indigenous women, that role alone cannot account for the multiple forms of racialized violence committed against Indigenous women in the area of study, or in Canada more generally. In the next section, I draw upon research participants' narratives to discuss both the gender violence of attacks on subsistence and the multiple sites and strategies of Indigenous women's resistance.

III. Diamond Mines, Social Reproduction, and Subsistence: A New Orientation?

Northern Indigenous women's experiences of the rupture of and resistance to the diamond mines are set against the backdrop of the racialized tension between place-based subsistence and time-based capital accumulation, which has intensified under the diamond-mining regime. This tension between time and place is more than a theoretical expression: research participants often expressed it as a pressing concern. Annie, a Dene woman and community leader and worker, compared Indigenous people's long-standing commitment to past and future generations – an ethic and way of being that she connected to her community and culture and held as primary – to the temporariness of the diamond mines and, in particular, to the environmental destruction of the mines, which she perceived as being conducted with limited forethought and for immediate profit. Annie described the diamonds simply as "rocks" devoid of nutrients, and she mourned the displacement of caribou (the primary traditional food source for the Dene) and the rupture in familial relationships for the sake of this empty signifier. Annie said:

I know the diamond mines, they say the water's good. They say. But I know that they use up a lot of water. And seeing the hole where they dig and dig. And all for a piece of rock. It's just a piece of rock. You're not going to eat a rock. It's not going to feed you. At least the land is there to have caribou, wildlife to be fed on.

I know they say they need money to function, to work. But, you know, it's our land. I remember the elders used to say, long ago, the elders used to say that the people from the south will come and start bothering you. And, exactly. It's happening. They're going to bother you about your land. They always have a media person who will say the water's good, bla bla bla. It's just, you really have to, the people who are in positions of power, like if you think of your young generation down in the future, and to stand up for them …

You have a table, you have a chair. You can replace that. But not human beings. When you have a relationship with your wife or husband and your children, it's very hard to replace that. You know, I hope somebody's listening to me. They can replace a lamp, their wood or that. And diamond mines, it's just a piece of stone. I think as human beings, we're really worth it to each other. And important. You cannot replace that. (Interview 207)

In comparing temporary material goods with long-standing human and land-based relationships, Annie illustrated the ways in which the temporary mining regime has undermined the place-based orientation toward people and land. Angus, like Annie, articulated the tension between the fixity of place-based social relations and the inherent temporariness, or detachability, of the diamond-mining regime's investment in the NWT as an extractive site:

And the past presidents of BHP and Diavik, they came and they shook hands and looked pretty and made promises. And where are they now? They're probably somewhere in a big house and, you know, your backyard is not our backyard. This is our backyard. So industry can come and say this or say that, but are they going to keep their promises? Where are the people who are actually the managers at the time? They're not here anymore. (Talking Circle One)

For Annie and Angus, their relationship to the region calls for a multi-generational approach that weaves together past, present, and future while acknowledging the ways in which the diamond mines are impacting the social reproduction of their communities.

For Annie and Angus, it is not the simple presence of capitalist production in the form of diamond mining that is of concern. Capitalist production is now part of the fabric of the northern mixed economy, and in its general form it is neither new nor (for many) unwelcome. Rather, the concern is with the particular *form* of diamond mining – the intensity of its temporal imperative – and its impact on Indigenous social relations. Temporariness was a common theme in both talking circles and interviews. Sarah, a Dene woman living in a smaller community, remembered promises made in the 1990s that were not honoured. Echoing Angus's critique, she said, "But this president, in the mid-nineties, with BHP Diavik, they're gone now. They're not living here anymore. Where are they living? They were hired, they were paid good money to influence the people, look good, shake hands, and now they're gone" (Talking Circle One). A number of participants compared this lack of attachment to the land – and therefore, a lack of consequences for its destruction or disruption – to their own rootedness. Elsie, another Dene woman, emphasized the violence of this tension: "And once the company and the business is gone, the negative impact falls and falls hard. And you see suicides. You see family break-up. You see communities empty out, and then what do you fall back on?" (Talking Circle 1).

While many research participants noted how useful the income of the diamond mines could be for general household expenses, most expressed concern about how the mines were shifting relations of production away from subsistence toward capitalist production. The urban location of this research likely accounts for the elevated concern as compared to research based in rural NWT locations (e.g., Harnum et al. 2014). However, it is also worth considering that women's stories have a unique ability to reveal tensions that have been obscured by the diamond mines' dominant narratives, which draw upon individual men's subsistence activities as a marker of healthy Indigenous/ settler relations (GNWT 2015). The abrupt and intense impact of the FIFO diamond-mining regime on a subsistence orientation amounts to a structural violence against community and intergenerational relationships – violence that manifests itself on both the land and the body. This is a violence that – due to the tendencies for the FIFO diamond-mining regime to introduce, or, in some instances, intensify, Western male-breadwinner relations in households, and in the context of long histories of gendered settler colonial violence – is experienced most severely by Indigenous women. At the same time, this is a structural violence that is met with both resilience and resistance enacted through community and kin relations; intergenerational education; and relations to the land, which I discuss, in turn.

a) Sharing, Community, and Kin

Inter-household relations are a primary site of community-level activities sustaining a subsistence orientation. Inter-household production and distribution has a long history and a strong presence in northern Indigenous communities (Dombrowski et al. 2013; Abele 2006; Usher et al. 2003) and distinguishes social relations in the mixed economy from the primarily nuclear relations found in most Western capitalist societies. These sharing relations have characterized some of the ways in which communities have approached their relationships to the diamond mines. For example, when negotiating bilateral agreements with the diamond mines, the Tłı̨chǫ upheld the value they place on community sharing by ensuring that benefits from the diamond mines would be distributed throughout the community, not just to those participating in the mines. Gibson MacDonald and colleagues (2014) offer examples of the ways in which the Tłı̨chǫ have used IBA funds to promote well-being and subsistence activities throughout their community. They write,

> In 2011 the Tłı̨chǫ Government allocated IBA funds to the promotion of harvesting, providing a harvester subsidy to each Tłı̨chǫ citizen at a time of the decline of a key species, caribou. This was a time of extreme hardship for the Tłı̨chǫ people, especially those in remote communities as they depend almost exclusively on caribou meat for their diet. In another example of the use of funds to support self-determination, the Tłı̨chǫ Government supported the Imbe Program with IBA derived funds. This program provides full-time summer jobs to 30 students who are pursuing their education in Southern Canada. These students study with elders and learn their native languages, how to set nets, dry fish and how to build canoe paddles. They travel the storied landscape with the elders and learn their Dene laws. As they hear the stories, they learn about their history of exchange and reciprocity. (70)

The Tłı̨chǫ have used diamond funding for clear tangible benefit. However, Gibson and colleagues (2014) also note how the sharing of diamond revenues has been impeded by lapses on the part of the diamond companies as well as by (I would argue), more generally, the inequitable distribution of revenue resources upheld through settler colonial resource law (see chapter 5). Furthermore, since the advent of the diamond mines, sharing relationships across households have been impacted by the absence of community members due to mine employment as well as by new inequalities within communities. Research participants described inequalities between mining and non-mining households as both material (e.g., a household with a big truck and many "toys," such as a boat or snowmobile, compared with households in communities that were struggling to meet their daily needs) and in terms of who shoulders what community labours and responsibilities. However, throughout the Canadian state's persistent attempts to restructure Indigenous processes of social reproduction away from subsistence, and the pressure to do the same imposed by the diamond-mining regime, kin and community networks have remained a powerful space of social reproduction. In interviews, Indigenous women consistently described how their extended kin and community networks worked together in times of strain and made it possible to pursue their desired goals. Indeed, kin networks created an important space of resilience and strength through processes separating subsistence from social reproduction. Sharing can involve material resources or labour, and research participants most frequently described the ways in which their extended family had provided much-needed labour to fill in gaps created by the demands of the FIFO diamond-mining regime. Women identified grandparents, in particular, as people who had taken up significant caregiving responsibilities while parents worked at the diamond mines. I asked Alica about the child care role that extended families take up:

> I think it goes both ways. Because, you know, obviously, when you're young, those strong family ties. But also a lot of strain on resources for family members. Like grandparents, for instance, I think are being asked to play a really significant role for a lot longer. Really play the role of raising the children. It's hard to say, but with my sister, for instance, the two weeks that her boyfriend was gone [at the diamond mine], she was working and we had to help her. So, for instance, she worked at the [place of work] and she had to be at work at 5:30. So she would drop off the kids every morning at 5 in the morning to my dad. And a lot of days we'd have to pick them up at school when she was working. So it really impacted her ability to work, not having that second person. (Interview 119)

These networks are a source of material and emotional support. They distinguish the experiences of many local Indigenous women from the experiences of partners of mine workers who moved from elsewhere to the NWT. For example, Christina, an immigrant from East Africa whose husband worked in the diamond mines, described feelings of extreme isolation when she followed her husband up to Yellowknife. For her, the experience of her husband's two-week departures to the mine was magnified by language barriers and by the work it took to build a new community. At the time of the interview, Christina indicated that she was building relationships with Yellowknife's immigrant community. Yet her story and her particular challenges were a reminder of the implications of the very distinct social locations and geographic histories of the women in the mixed economy in and around Yellowknife. As one community worker noted (Interview 303), the struggles of socially reproducing Indigenous communities in the mixed economy are at least acknowledged (however challenging and sometimes violent they are), whereas the challenges of social reproduction for other minority communities remain, for the most part, invisibilized.

By contrast, Indigenous women who were linked to local kin and community networks spoke of the resilience, strength, and happiness that flowed from these social spaces and the ways in which they drew upon kin and community supports to manage the added responsibilities and loneliness they experienced while their partners were at the mine (recalling, here, the stories of Doris and Rose and their extended communities in the previous chapter). The "resilience" of particular populations is sometimes deployed as a depoliticized justification for social restructuring.[9] Conversely, I use the term to point to the strength of these communities and the persistence of their unique social formations, not to diminish the difficulties, the burdens, and the violence that have accompanied the restructuring of departments of production.

Resilience, however, should not be confused with resistance. Resistance to capitalist restructuring was also demonstrated, through the sharing of labour and resources within community and kin networks. More specifically, research participants described ways in which they privileged the social reproduction of their community and kin networks over the demands of capital. Iris, Doris, and Liza all left work in the diamond mines to prioritize their inter-household care relations (looking after nieces and nephews, grandchildren, and vulnerable community members). Other Indigenous women participants – like Angela and Rose – explained that they had never sought work in the diamond mines in part because maintaining their community networks was too important to them. In the context of a mixed economy in which there is an ongoing struggle between capitalist, state-sanctioned modes of social reproduction and place-based, Indigenous modes of social reproduction, extended networks of intimacy and care are a space where the latter can be supported and nurtured outside of the confines of the former. The women I interviewed described extended networks as a source of joy, stability, and strength, but also as spaces of responsibility and labour – labour that often conflicted with their wage-labour roles, most notably in the diamond mines. In interviews, it was just as common for women who had worked at the diamond mines to describe their community and kin-level work as conflicting with their work at the diamond mines as it was for them to discuss labour involving their own children.

By making the intergenerational social reproduction of their communities their primary concern, Indigenous women in and around Yellowknife – in their relationships and their labour – were challenging the totality of capital as well as maintaining a continuity of a subsistence orientation that values place-based practices of care and intimate relations to people and places over the atomized separateness so conducive to the neoliberal extractive order. However, while it is important to celebrate the decolonizing strength acquired through the sharing of community and kin networks, one must not diminish the difficult and sometimes constraining roles Indigenous women take up within these networks. Recall Iris's discussion in chapter 6 of the difficulty she had retaining a job at the mine because of her responsibilities to her extended kin networks. Iris described her extended kin and community as her core responsibility. She saw these relationships as a source of strength, but she also noted the ways in which her heavy responsibilities had made life more difficult for her. While Indigenous women's orientation toward the social reproduction of their kin and community is clearly an act of resistance to the totalizing impulses of capital and the Canadian state, simultaneously and contradictorily their labour is structured through patriarchal capitalist male-breadwinner ideologies and materialities (imposed by the Canadian state and intensified by the diamond-mining regime), which make women responsible for unpaid and often invisibilized or devalued labour outside of the site of capitalist production. Thus, while community and kin networks are a site of resilience and resistance to the feminizing and "nuclearizing" of social reproduction, these networks are not themselves exempt from these processes of restructuring. As Hilary, a community worker, said:

> The expectation is that if anything tough happens in the community, that the women have to be there. Like, if there's a sick parent or an older person, the men in the community are really not expected to take on the extra roles. Women need to take them on. So that makes it really difficult for women [to work for the diamond mines]. (Interview 201: 2014)

Thus, newly intensified gender inequality in kin and community responsibilities accompanies the new inequality between households.

Research participants identified new inter-household inequalities as a barrier to community sharing. For example, some discussed how mine workers were resented by and felt pressured by other community members. Alica, who had worked at a diamond mine, described a co-worker's experience in this way:

> So when he started working there [at the mine] – and he also had a really hard time taking the job for the same reasons I did, he felt like a traitor, but he's worked there since they opened – his family started to have money. So they started to buy nice things, like skidoos and trucks. Because they could afford to. But he found there was a lot of jealousy in the community, so he stopped. It caused a lot of jealousy between the haves and the have-nots. (Interview 119)

It often happens that mine workers have more financial resources than non-mine workers but at the same time are less available to take part in community activities and labours (Interviews 207, 303). Because most mine workers are men, this exacerbates community and familial tendencies to expect women to undertake community labours. At the same time, mine workers are targeted for major loans and purchases, such as home mortgages. These financial commitments further draw their individual and/or household labour and resources toward the re-production of nuclear households and away from community needs. To put it another way, while family and community may expect mine workers to contribute part of their pay to friends and family in need or to community resources, in line with long-standing values of community sharing, that pay may already be stretched thin by household debts. The resulting strain on community relations poses a challenge to the practices through which community knowledge and subsistence practices are relayed through generations.

b) Subsistence and Education

Intergenerational education is integrated into subsistence production in Indigenous communities in the region of study. Research participants described their experiences as both teachers and learners as integral to their identity, their relationships, and their knowledge and skills. Largely as the result of the structural violence of residential schools, Indigenous communities are faced with the challenge of simultaneously mending intergenerational ties and relying on those ties to transmit knowledge and build well-being, thereby sustaining their communities across generations. In a talking circle, Elsie spoke of the generational gap in learning as a result of residential school: "I don't know my language. I don't know how to hunt and trap or live off the land. My children went to school in the Beaufort Delta to learn their language. To learn what people do on the land and everything. I grew up in a residential school" (Talking Circle One). Debbie, a Métis woman in the same talking circle, described the impact of these restructuring processes on community education: "There were teachers in our community that helped the other ones. That's gone now. I don't know why. But more and more we're losing our traditional teachers who can prepare the other ones to be healthy" (Talking Circle One).

In the face of these challenges, a number of research participants expressed concern about the diamond mine work that separated young people from opportunities to participate in subsistence production and thereby learn from older generations. Angela described the challenge her family members faced in this regard:

I look at my uncles that live in Rae [Behchokǫ̀] and they're very strong with their traditions. They're hunters, they're fishers, they're gatherers. And they really invest their time. But their sons, my cousins, they don't have that. Because when they come home [from the mines], they've missed out on so much with their family life and social life that they don't want to go hunting. And they're like, whatever, my dad did it or my uncle did it, so I don't have to worry about it.

So I feel like there's going to be this huge gap. Whereas my cousins' kids are still going out with uncles and grandparents to do that. But for how long? And I can see it. A lot of my cousins, as soon as they're done work, they go home, if they're not fleeing … it's like, let's go to Yellowknife. Let's shop. I see a lot of my cousins, as soon as they get home, they feel like maybe the money will make some happiness in their home. Where there's a lot of items that, if they were to be living amongst their family, or working and living amongst their family side by side, they wouldn't go and spend that much money. (Interview 303)

For Angela, work in the diamond mines impeded participation in subsistence education. Her insight speaks not just to the geographic separation and the time away but also to the new ideologies and material imperatives that accompany the work.

A number of research participants discussed the incentives to direct one's education or training toward the needs of the diamond mines. The concerns were not just that diamond-mining work would keep young people away from subsistence education, but also that diamond-mining work would keep young people away from non-extractive-related schooling and experiences. Elsie worried that adolescents would say, "Why do we bother going to school when we could make all this money?" (Talking Circle One). For Amy, a recent high school graduate in Yellowknife, the concern was less that students would not finish high school and more that they would be convinced to orient their post-secondary education and work experience toward the needs of the diamond mines. Discussing her last year of high school, Amy listed job fairs, co-ops, and substantial scholarships as ways that students were being incentivized to consider the mining industry:

> People are shoving things down your throat and the one that they shove the hardest is diamond mines. And come back to the North. And the way that they see that being achieved is these sciences. And if you're not willing or not able to get a job in, you know, an academic field, do the university route, they really push the skilled trades. (Interview 104)

Amy's experience echoes Little's (2007) argument that masculinized mine labour is valued more highly than feminized social and health labour around Yellowknife (31). Recall, for example, the defunding of the NWT Aurora College social work program at the same time the territory began funding a new $10 million mine training centre at the college. The gendered implications of the focus on diamond mining at the expense of other feminized forms of capitalist production are a reminder that in the mixed economy, not all capitalist production is equal, nor is it equally gendered or similarly related to Indigenous social relations. Indeed, the devaluation of feminized work in relation to the diamond mines cuts across capitalist production, subsistence production, and social reproduction – an important continuity, given that people do not experience or labour in these departments discretely.

But at the same time, the pressures the diamond mines are placing on intergenerational education are being challenged at multiple sites through decolonizing processes of intergenerational knowledge transmission, from the incorporation of traditional knowledge into formal education, to the nurturing of informal kin and community networks, to knowledge transmission at the mines themselves. Indeed, just as Kakfwi (1977) noted that, in bringing Indigenous people together under the same roof, residential schools unintentionally served as sites for building pan-NWT Indigenous resistance, a number of research participants described the ways in which mine workers used the mine site as a space of knowledge transmission. For example, as Nahanni writes (1992), in Dene tradition, language is considered fundamental to intergenerational learning and the protection of knowledge. Like many Indigenous languages, the Tłı̨chǫ language holds within its names long-standing community knowledge and stories. Beth described the importance of the Tłı̨chǫ language in her own family, proudly explaining that her husband had found a group of men with whom to speak Tłı̨chǫ at the mine camp. Iris, for her part, discussed collecting herbs and plants at the mine site, and spoke of the importance of finding good Inuit leaders at camp (Interview 101) to take workers out on the land.

Outside of the mine site, Indigenous women discussed multiple strategies for protecting and pursuing Indigenous intergenerational education. Earlier, I shared Liza's experience of joining a group of Indigenous women who gathered plants and taught one another traditional medicines. Liza's experience is a distinctive story of knowledge transmission and decolonizing resistance: she is not originally from the NWT, and in fact left her home Indigenous community in another part of Canada because she and her husband were having trouble finding work, and they were told he could get a job at the diamond mines. For Liza, finding a group of Indigenous women who were committed to using and sharing their traditional knowledge was a source of strength and joy at a time of change and loneliness (Interview 207). Tory, whose story was also shared earlier, was one of the younger women who participated in traditional activities with Liza and her friends. Like Liza and Doris, Tory has used the relationships and the knowledge shared by older members of her kin and community to find strength against the racial and gender violence she has experienced, and to find ways of making a living outside of the diamond-mining regime (Interview 116). Thus, in tandem with First Nations government programs designed to build intergenerational knowledge transmission, research participants offered informal and community-based methods through which past and present ruptures in intergenerational relations are mitigated and mended.

c) The Land and the Body

Across interviews and talking circles, discussion of the relationship between the diamond mines and subsistence was grounded in an attention to the links between the land and the body, expressed through concerns both embodied and structural. Indeed, violence to the land is tied up with an immediate violence done to the health and well-being of bodies, as well as a structural violence to a subsistence orientation. In Annie's critique of the diamond mines, she expressed the environmental impact of the mines through an ontology of place "as a way of knowing, experiencing, and relating to the world" (Coulthard 2010: 79). Similarly, when I asked questions in interviews and talking circles about the diamond mines – as sites of physical displacement and environmental destruction – many women began their response with a discussion of their children, responses influenced by broad concerns about the ways in which resource extraction will impact future generations, but also driven by contemporary material proximities between day-to-day and intergenerational social reproduction and the land. For example, a number of women brought up concerns for the caribou, traditional meat for the Dene, and their shifting migration patterns and shrinking numbers as a result of the mines. Sarah put it this way:

> The diamond mines also changed the caribou migration. And because they changed the caribou migration, the caribou don't go as far as they used to. They don't go where they used to because their habitat is being taken over. Their numbers are obviously decreasing. And so that is impacting my family, my community, my culture. Where a lot of us don't have access to caribou, our traditional meat. My daughter, who's now twenty-one, she just found out she's diabetic. She's borderline diabetic. She's not obese, she exercises a lot, and when we found that out, we were surprised.[10] We said, how could that be? But that's because our food and our cultural way of living has changed. Our food has changed. Our diet has changed. So, she has to eat a strict diet now and that's just how it is. (Talking Circle One)

Sarah's concern articulates a relational approach to the land: rather than environmental destruction being an *object* of concern – something separate or alienable – she links caribou herd disruption with the health of her daughter, an expression of place-based interdependence. Debbie held the same concern, and told this story of resistance, which got a good laugh from the group:

> And I no longer have a food source [as a result of the diamond mines]. I remember one time we were out hunting and we used the [mine] road. We were grateful for the road. But right in front of us, a group of hunters that went out and, honest to god, it was just like Vietnam. Boom, boom, boom, boom, boom. And this whole herd of caribou was hunted, slaughtered. And this group of hunters was standing around, not knowing what to do now. Because they'd never hunted before. But they see an animal and they kill it. So my husband jumped out and showed them how to harvest the animal. And he took all the delicacies, and said, "oh, this part's no good." [laughs] So we benefited, thankfully. But, you know, lots of times we went out on that road and people aren't even seeing caribou anymore. (Talking Circle One)

The Indigenous communities of the North share a concern for caribou. In response, the Tłı̨chǫ government has initiated Boots on the Ground, a caribou-monitoring program based on the traditional knowledge of Tłı̨chǫ and Inuit elders and harvesters (Dedats'eetaa 2019). The program has reported caribou herd decline since Ekati mine was established, beginning with herd disruption as a result of diamond prospecting. It writes that "main migration routes were from Ekati, to Gordon lake, to Yellowknife was a main migration route. Now the mine and winter road has taken over the caribou migration areas" (92).

Western-style environmental assessments (whether sanctioned by the diamond industry or not) characterize the diamond mines as relatively clean[11] – though I note that the bar for "clean" extraction is incredibly low. However, given that the diamond mines are being compared to a century of extractive projects that dumped contaminants onto the Subarctic and Arctic land with little or no thought of remediation, the displacement of animals and the unknown levels of damage to the earth were major environmental concerns for the research participants. These Dene, Métis, and Inuit women viewed the health of the land as directly related to the health and well-being of the people, in the material sense of traditional food sources but also in emotive, interpersonal, and cultural ways (recall, here, the porous body that shapes and is shaped by the social relations it inhabits, which include relations to the land). Alica, whose grandfather's trap line once ran through the land where a diamond pit has now been sunk, explained that she was physically ill when she visited the diamond mines (Interview 119). A number of research participants discussed the meaning their traditional land held for them and the role land plays in intergenerational social reproduction. When I asked Rose about bringing her child to visit family who lived in small communities, she related spiritual, emotional, and interpersonal health with visiting the land to which her family is connected:

> With the land, you hear Aboriginal people, and for me, talking about our relationship with the land. I feel like it's something that I'm not even fully aware of. My soul, unconscious, my body, anytime I go see my aunty and my uncle at their camp or do any kind of cultural activities, my connection to the land, it's not even something that's in my mind, like the forefront of my mind. It's just such a release there. (Interview 102)

The relationship between land, subsistence, and the social reproduction of Indigenous communities is a thread wound tightly around different times. Indigenous communities, particularly the Yellowknives Dene, have not forgotten the devastating environmental impact of the gold mines, or the illness and death in their community as a result of the arsenic left by those mines. And indeed, as Indigenous people engage in the mixed economy, the arsenic deposits in the land on which they labour, though literally frozen in permafrost and metaphorically frozen in a politics of silence,[12] continue to pose a threat to the health of the people, land, and animals. Reflecting on this history, a number of research participants pointed to the environmental destruction wrought by extraction in relation to their concerns about cancer. For example, Beth linked her concern for the caribou herds with her observations of cancer rates in her own community and in Indigenous communities across the country:

> For me, I don't really like the idea of the mines being out here. Like, I love our land. I love our water. I just feel like I don't want them here. I don't care how much money and jobs it brings for our community. I just don't like the fact that they may be contaminating our land, our water, and our animals. Because we still live off of our traditional foods. I'm afraid that, even though they do monitoring and testing and things like that, down the road we may find that some of the animals are contaminated and maybe that it will possibly lead to some of the reasons that people are dying from cancer. We've lost a lot of people to cancer. And they're still finding people with cancer and that really scares me. And I'm not really happy with the mining companies being here … I love my traditional food. But I'm just also scared. I don't want it to be contaminated. (Interview 107)

The research participants' accounts about their relationship to the land stimulate an analysis of the ways in which the corporeal fits into the structural. Sarah's daughter's diabetes, which she attributes to the loss of her traditional food source, Alica's pain at the destruction of her family's traditional trap line, Beth's concern about cancer … these articulations transcend the distinction between structural violence and embodied violence. Furthermore, all of these concerns are tied to the intergenerational social reproduction of Indigenous communities: a grounding of past, present, and future *in place*. These violent relations are obscured by government and industry approaches to the land. Specifically, government and industry reporting has separated the diamond mines' environmental impact from their social impact (Caine and Krogman 2010: 78). This is an act of settler erasure of the *social* relation to the land.

Conclusion

There is a fluidity in the ways in which northern Indigenous women expressed their relationships to the land, their relationships to their children and loved ones, the systems of meaning that threaded through all of these relationships, and the violence they had experienced structurally and corporeally. Concerns and experiences related to social reproduction, subsistence, and the land were not articulated discretely. The participants articulated the ways in which subsistence has been impacted by the diamond mines in narratives weaving together experiences of interpersonal violence and shifts in social relations at community, household, and interpersonal levels.

How, then, in relation to the FIFO diamond-mining regime, have the gradations in orientation between capitalist production and subsistence production changed the day-to-day and intergenerational social reproduction performed by Indigenous women living in and around Yellowknife? Do the diamond mines represent a reorientation away from subsistence and toward capitalist production, and if so, what is the racialized and gendered quality of this shift? What does resistance to this shift look like? Taken together, chapters 6, 7, and 8 have elevated Indigenous women's labour – subsistence and social reproduction – as a site of colonizing and decolonizing struggle, of violence and creative resistance. In the Yellowknife region, social reproduction commitments toward extended kin and community are tied up in an orientation away from the demands of capital and toward the daily and intergenerational social reproduction of Indigenous community social relations. This orientation has been strained by the tempo and temporariness of the diamond-mining regime and its patriarchal ideology, which has imposed nuclear, male-breadwinner/female-caregiver relations on relations of social reproduction that extend far beyond the imagined picket fence.

Capitalism, in the abstract, has a totalizing impulse; concretely, the diamond-mining regime and its FIFO structure, combined with its targeting of Indigenous workers, has brought new pressures on the mixed economy in and around Yellowknife. These pressures have resulted to some degree in a restructuring of social reproduction and subsistence. Faced with these pressures, research participants exemplified the day-to-day labours oriented not toward the demands of the Canadian state or the imperatives of capital, but rather toward the social reproduction of their community. Just as much as large-scale battles for land, the often-intertwined daily labours of social reproduction and subsistence – women showing their daughters and nieces the plants that are medicine, teaching children the hunting paths of their ancestors, fishing, cooking, sewing, and drying meat – at once intimate and transgressive, sustain the distinct, land-based ways of being and knowing of the northern mixed economy.

9 Conclusion

Speaking as a grandmother myself, diamonds, first it's oil, now it's diamond mines. I think of the damage that's being done to our sacred ground. We're lucky to be able to sit here today and have some choices, but when I think of the future generations, our great-grandchildren and their future generations, are they going to be able to have clear water? Even the fish and all the environmental hazards that are related to diamond mining. Water is becoming a commodity. So I really fear for the lives yet to come for our future generations. I always think about the long-term effects of what they're doing to our sacred ground.

And someday we should be able to realize that money cannot be eaten. Because in the long run, money's not going to mean nothing. For them to be able to sustain and be able to live off what we have left in our land, for me, I find it so shocking that we use these as a fast way to get what we want here. And don't really think about the long run, about the future generations, about our children and grandchildren.

– Talking Circle One

Ekati, Dominion, and Gatcho Kué diamond mines are some of the largest open-pit mines in the world, carving gaping holes into the Arctic land. While the vast landscape obscures its size, up close, the heavy equipment used to operate the mines looks as though it was imported from a land of giants, to displace raw materials on an otherworldly scale. These processes of geological displacement – imposing as they are – operate on a scale that can obscure the day-to-day forms of *social* rupture on which the diamond mines rely. Space and time are central – and contradictory – aspects of FIFO extractive regimes. FIFO extraction, with its emphasis on variable over fixed capital, its resort to

Image 4. Diavik Diamond Mine (Photo courtesy of Rio Tinto Diavik Diamond Mines. Photo by Pat Kane).

chartered flights rather than mining towns, and its intensely temporary nature, is in some ways a perfect example of capital's tendency to annex space with time. That is, in the pursuit of new processes of (temporal) accumulation, FIFO structures production in ways that circumvent place-based constraints to whatever extent possible. However, just as resource extraction annexes space for temporal and temporary valuation, so too is it intrinsically reliant on space. As Rainie and colleagues (2014) write, "a dependency on natural production limits spatial flexibility ... It is difficult to move a mine, no matter how much a particular company may wish to do this" (104). There is, then, a double contradiction to a form of capitalist production that is rooted both in materiality and in the extraction of that materiality in the context of a mixed economy grounded in the reproduction of place-based relations. The gendered processes of the inherent rupture of these contradictions – of displacement and restructuring – within the social relations of the mixed economy of the Yellowknife region and the violent attributes of these processes have been the subject of this book. Rather than a fixed "mixture" of Indigenous and settler modes of life, I have approached the mixed economy as a dynamic site of settler colonialism, decolonizing resistance, and day-to-day multi-scalar negotiations of these differing imperatives.

My aim has been to examine the ways in which shifts in departments of production instituted through the diamond-mining regime play out in the lives, labour, and bodies of Dene, Métis, and Inuit women living in and around Yellowknife. In developing an expanded conception of production as my analytical framework, I have approached capitalist production, social reproduction, and subsistence as departments of production at the same level of analysis within the mixed economy of the Yellowknife region. The mixed economy is characterized by tension between the temporal imperatives of capitalist production and the place-based imperatives of subsistence. However, beyond this tension, the mixed economy has been shaped through the land-based modes of life (Coulthard 2014) of northern Indigenous peoples, through the processes of settler colonialism that brought new people and new modes of production to the North, and by the activist presence of the Canadian state. Indeed, the tension in the region's mixed economy between place and space is both structured and mediated by the past and present role of the Canadian state in shaping social relations in that economy. I have emphasized the de/colonizing implications of shifts in relations between departments of production, arguing that it is not enough to note that the mixed economy is comprised of three departments of production; we must also trace the relations of power between these departments as they play out through the racialized and gendered social relations of the region. In this way, FIFO diamond-mining emerges as not just another form of capitalist production, but as a specific regime that has intensified the capitalist emphasis on temporary modes of accumulation and the racialized and gendered tensions between departments of production.

The diamond-mining regime sits in continuity with the global violence of resource extraction, embedded in the racializing and gendering processes of colonialism and imperialism. The modern diamond industry was established as part of the British imperial project in South Africa and came to fund imperial expansion on the African continent. However, as happens so often with luxury commodities, from the twentieth century onward, the diamond industry, tightly controlled by De Beers, worked hard to distance diamonds from their violent relations of production. Diamonds became synonymous with Western romantic love, aspirational gems, precious yet accessible to the middle-class consumer. This imagery was in danger of being washed away by "diamond conflicts" in Sierra Leone and Angola in the late 1990s and by the attention that Western NGOs began to pay to those conflicts. Around that time, Canada emerged as just what the diamond industry needed: a source of (ostensibly) conflict-free diamonds, devoid of the stain that threatened the industry's fragile imagery. The rhetoric of "responsible" Canadian diamonds has obscured the gendered and racialized violence that has long tied the northern diamond mines to extractive projects across time and place. When we expand our analysis beyond the site of production, and emphasize Indigenous women's experiences and la-bours, these violent continuities come to light.

In particular, I have focused on social reproduction as a site of de/colonizing tension. Indigenous women's activities associated with social reproduction sit at the locus between capitalist and subsistence production, as social reproduction is required for both and operates rela-tionally with both. Simultaneously, then, social reproduction is a site of potentially violent tension and decolonizing struggles to maintain and cultivate place-based Indigenous social relations. I asked what processes of social change, what colonial impositions, what creative resistance, and what violence is made evident when social reproduction is placed at the centre of analysis. In what follows, I reflect on these questions, discussing the insights that have emerged from the preceding chapters and identify-ing emerging questions and areas of research. I begin by synthesizing the analysis of the three departments of production and how their relations have shifted since the advent of the diamond-mining regime. I then turn to the question of settler colonial violence, reviewing the insights that have emerged in this research and bringing these into conversation with the broader violence against Indigenous women in Canada. Given that the life of the NWT diamond mines will likely be relatively brief, I end with a question about northern life after diamonds, locating the diamond industry in the broader northern history of rupture and resilience.

I. Shifting Relations of Production

a) Capitalist Production

In the Yellowknife region, diamond mining carries far more weight in economic discussions (from kitchen tables to the legislative assembly to the towers housing Indigenous and Northern Affairs so many kilo-metres south in Ottawa) than its employment numbers would suggest. For example, in 2017, just 3.5 per cent of the NWT population was em-ployed in the diamond mines (GNWT 2019a). Notwithstanding the diversity of both waged and non-waged labour in the Yellowknife re-gion, many research participants observed the ubiquity of the impact of resource extraction across time. Jenny put it this way: "I didn't even want to work for the mines, but I soon learned that it doesn't matter who you work for. You're working for the diamond mines anyway" (Interview 115). The gold-mining regime had introduced a "temporary permanence" for the region's settlers, and that regime established Yellowknife as a mining town, thereby investing in fixed capital. Settler men took up stable jobs with strong union protections. This model was not to last. Indeed, the shift from the gold-mining regime to the diamond-mining regime was largely a temporal one: the presumed settler stability of the gold-mining regime bumped up against a new approach to temporariness, tempo, and place in the form of the diamond mines. The diamond-mining regime followed contemporary extractive trends (Peck 2013) and the increased temporariness of neoliberal processes of production. Thus, it embraced a FIFO model of development, a model that intensifies the temporariness of resource extraction with its emphasis on variable over fixed capital. The diamond-mining regime's intensified temporariness was evident both in the FIFO structure itself – workers' presence at camps was on-again/off-again – and in the regime's neoliberal approach to employment, which involved short-term contracts, subcontracting, and a limited union presence.

The diamond mines are also distinguished from their predecessors by a new approach to engagement with the northern Indigenous population. The Canadian state took an activist role in restructuring Indigenous social reproduction and subsistence in the mid-twentieth century; but at that time, the Canadian state and private capital made only limited efforts to integrate local Indigenous people into capitalist production. The gold mines had been established on Dene land with no consultation with Dene people. Northern Indigenous workers were relied on by the gold mines for their labour and their land-based expertise, but only limited attempts were made to employ them as full-time mine workers. Conversely, the particular adaptation of the global FIFO trend by the NWT diamond mines has been shaped by the Indigenous movements of the late twentieth century, which won new practices in Indigenous consultation and participation (Bielawski 2003; Gibson 2008; Irlbacher-Fox 2009). The diamond-mining regime has responded to the sociopolitical imperative to engage the Indigenous peoples of the region with an approach oriented toward participation by Indigenous communities as stakeholders and Indigenous individuals as potential business operators and workers, though it has paid little heed to Indigenous peoples's distinct histories and relations with the land. The diamond-mining regime's approach to Indigenous communities is in fact a contradiction: northern Indigenous peoples have been approached as "stakeholders" because of their acknowledged relationship to the land, yet this acknowledgment denies the antithetical relationship between Indigenous land-based modes of life and the imperatives of extraction. Instead, the diamond-mining regime has approached Indigenous peoples primarily as an identity category to be targeted for employment and subcontracting, and this has externalized the contradictions between the diamond-mining regime and social reproduction and subsistence (see below). The agreements made between the diamond mines and the territory (SEAs) and the diamond mines and the Indigenous communities (IBAs) are attempts to rhetorically resolve the tension between extraction and Indigeneity, as a racialized subjectivity marked by a relationship to place. I argue, however, that ultimately the diamond-mining regime depends on a dematerialized interpretation of "Indigeneity" as a racial category, as a means to obscure extraction as a site of contradiction between fixity and temporariness, particularly temporariness as embodied in the physical and social restructuring of earth and people.

Furthermore, like other extractive models, the diamond-mining extractive regime is structured through a settler ideology that masculinizes capitalist production, feminizes and naturalizes social reproduction, and obscures subsistence altogether. The FIFO structure, which separates capitalist production from social reproduction, has raised obstacles to women's full participation in wage labour at the mines, for it has made women responsible for social reproduction in the community and home and created a labour hierarchy at the mine site, wherein mostly Indigenous women perform on-site labour for the daily maintenance of mine workers in the context of a hypermasculine work environment. Indigenous women's participation and non-participation in the diamond-mining regime is located at the racialized and gendered interface between departments of production, which sometimes manifests itself as experiences of embodied violence at the mine site and in the women's home communities.

Thus, while in some ways the diamond-mining regime carries on the history of settler-driven resource extraction around Yellowknife, its FIFO tempo and its new approach to Indigenous relations have placed novel pressures on Indigenous women's social reproduction and subsistence labour.

b) Social Reproduction

Settler colonialism is a form of colonialism wherein settlers claim autonomy and sovereignty over a specific space and, as noted by Veracini (2010), also claim a regenerative capacity and right. This "regenerative" impulse articulates in the Canadian context as the social reproduction of social relations informed by Western capitalist materiality and ideology; in the region of study, those social relations interact dialectically with the reproduction of Indigenous social relations. Social reproduction as a central site of settler colonial intervention helps explain the contemporary (structural and embodied) violence against Indigenous women in the area of study and in Canada more broadly. Like other racialized women in Canada (Thobani 2007), Indigenous women have been targeted by state surveillance and restructuring of their day-to-day and intergenerational social reproduction. Indeed, during the time of so-called separate development in the region of study (the gold-mining era), the Canadian state took an activist role in restructuring social reproduction activities performed by Indigenous women, notably through residential schools and the relocation of Indigenous communities to permanent settlements. Furthering twentieth-century Canadian state interventions cumulatively imposing a patriarchal nuclear family structure, including residential schools and the high rates of state apprehension of Indigenous children that followed, the imperatives of the diamond-mining regime have augmented pressures toward a male-breadwinner/female-caregiver nuclear model. Indeed, research participants emphasized the proximity between the mid-twentieth-century efforts of the Canadian state that targeted social reproduction and the contemporary demands on social reproduction made by the diamond-mining regime. Martha discussed the proximity of these two processes of rupture at the site of social reproduction:

> When you look at a population that's been so dedicated to preserving a natural environment for so long, and then everything gets shaken up. And then as we're all trying to reconnect with it, and it's like, residential schools are closed as of, like, twenty minutes ago. Now we have diamond mines. And I think it's just a lot of really big shocks. (Interview 202)

In this way, the former (state-driven) ruptures helped facilitate the latter (mining-driven) ruptures.

Thus, while it was not the diamond mines that introduced a male-breadwinner/female-caregiver model to the region, the FIFO structure and the expanded participation of Indigenous people in the extractive regime have combined to heighten the pressure for this. FIFO is a spatial articulation of the separation of capitalist production from social reproduction, and that separation is antithetical to subsistence-oriented relations in the mixed economy, wherein waged and unwaged and masculinized and feminized labour can be oriented toward the same goals of meeting the daily and intergenerational social needs of household and community. The extractive spatial and ideological requirement to adhere to the separation between capitalist production and social reproduction, then, has resulted in a new (or newly reinforced) feminization of the responsibilities for, and performance of, social reproduction. In this way, the FIFO diamond-mining regime has helped intensify Western patriarchal gender relations so that social reproduction is increasingly reoriented toward the demands of capital and away from subsistence. This intensification means that, through intensified and marketized household pressures (through, for example, the move to market housing, the physical separation of workers and households from home communities, and greater demands on household labour), social reproduction is being pulled toward the nuclear level at the expense of more fluid inter- and intra-household relationships. This is generating a rupture in place-based processes of social reproduction.

Recall, however, that according to Tłı chǫ cosmology, a rupture is not an end or a break in social relations. Rather, it is a reconfiguration. This concept is a reminder that the restructuring of social reproduction as a result of the FIFO diamond-mining regime has happened incompletely and in relation to the many and multi-scalar ways in which Indigenous women continue to orient their social reproduction toward land and community relations. Indeed, Indigenous women's social reproduction is a site of creation and agency rather than a site of "victimhood," or a site that is merely acted upon. In interviews and talking circles, research participants consistently described the ways in which social reproduction was taken up as a site of strength and community-building – labour that cannot, and should not, be approached discretely from contemporary subsistence production. In discussing the struggles of social reproduction at the community level, research participants described the innovative ways in which intergenerational subsistence education has persisted, even with the push to train young people for work in the diamond mines. And while the pressures to orient labour toward the nuclear level should not be minimized, participants described their extended kin and community networks as a source of strength. This strength can be read as resilience – as kin and community helping women manage the difficulties imposed by the diamond mines – and as a site of decolonizing resistance. Many of the women interviewed described prioritizing these networks – and their associated social reproduction – over the imperatives of the extractive regime, in this way reproducing an orientation toward communal needs, or, subsistence.

c) Subsistence Production

Recall that the division between social reproduction and subsistence emerged throughout the twentieth century as a result of Canadian state intervention and the imperatives of capital and was intensified by the FIFO diamond-mining regime. To account for the complex and inter-dependent ways in which Dene, Métis, and Inuit women engage in the three departments of production in the area around Yellowknife, I have drawn upon the notion of a "subsistence perspective" (Mies and Bennholdt-Thomsen 1999) as a means to assess the qualitative, subjec-tive, and relational shifts in orientation and power between and within departments of production. Learning from Coulthard (2014) and Coul-thard and Simpson's (2016) articulation of land-based practices, ethics and relations, I have approached subsistence, and its orientation away from capital, not as an anachronistic cultural practice or as recreation, but rather as a meaningful alternative within the contemporary north-ern mixed economy.

Subsistence production is at odds with the temporariness and tempo of the FIFO diamond-mining regime, which emphasizes capital accu-mulation at the expense of long-standing relations between people and the land. By approaching subsistence as an individualized recreation activity that diamond-mine workers can take up during their "time off," the extractive regime has rhetorically resolved the tension be-tween subsistence and extraction, thereby positioning itself as "Indige-nous friendly." However, this conceptualization obscures the relational nature of subsistence; for example, it fails to acknowledge that subsist-ence activities are, for the most part, undertaken communally and in line with the seasons, the behaviour of land and animals, and the needs of a community (rather than when a person happens to have time off); and that subsistence activities tie together activities on the land (for example, hunting a caribou) and in the community (all the labours that go into the preparing caribou for food and household and community materials). Given Indigenous women's roles in social reproduction and subsistence outside the site of the diamond mines, a focus on their so-cial relations exposes the tensions between the demands of the FIFO diamond-mining regime and day-to-day and intergenerational social reproduction oriented toward subsistence.

The social reproduction of inter-household linkages and subsist-ence education together demonstrate the role of community-level intergenerational social reproduction in decolonizing resistance. Re-search participants reported that, faced with the pressures of the FIFO diamond-mining regime, they had succeeded in carving out the space to socially reproduce their communities. At the mine site, Iris and her friends went out on the land to pick medicines, and Beth's husband and his friends spoke Tłı̨chǫ Yatıì. Doris used her sewing and bead-work to make a living outside of the mines, while Liza joined an Indig-enous women's group to share her knowledge of plants and medicines. Rose drew on her kin networks to leave a violent partner, and Tory drew her strength from the traditional teachings of her elders. In talk-ing circles, Annie and Angus spoke of post-extractive futures guided by community norms, relationships, and commitments – a vision for a future mixed economy made possible through ongoing transmis-sion of land-based ontologies, practices, and knowledge. These are specific findings that cannot be generalized; however, taken in com-bination with the demonstrable history of Indigenous resistance and contemporary engagement with subsistence production, the examples herein of reproducing the unique social relations of the mixed economy demonstrate the ways in which the totality of capital is resisted through day-to-day labours.

At the same time, a focus on Indigenous women's engagement in subsistence and social reproduction sheds light on the structural and embodied violence of the FIFO diamond-mining regime. The links be-tween the restructuring of subsistence and the well-being of the land and the body are profoundly racialized in gendered ways. This recalls for us the strength and sustenance that a subsistence orientation offers to many Indigenous women, as well as the vulnerabilities that orien-tation faces to the violence of the reorganization of land and labour to fulfil the imperatives of capital accumulation, a tension to which I now turn.

II. Gendering the Colonial Violence of Extraction

The appalling violence against Indigenous women in Canada demands that we pay attention to the relationship between embodied violence, the social reproduction of Indigenous social relations, and the pursuit of new spaces of accumulation under capitalist colonization. Work by Indigenous activists and scholars (Maracle 1988; Anderson and Law-rence, eds., 2003; Smith 2005; Simpson 2014; Simpson 2016), especially in the last three decades, has challenged racist colonial discourses and materialities that dismiss, minimize, or naturalize this violence. In Can-ada, national initiatives like the Sisters in Spirit inquiry, annual Straw-berry Ceremonies, and the National Day for Murdered and Missing Indigenous Women have pushed for approaches to addressing vio-lence that link past and present violent structural processes of settler colonialism with specific embodied instances of violence against Indigenous women.

The present research has approached the relationship between settler colonialism and violence against Indigenous women through the lens of Indigenous women's work in socially reproducing the place-based relations of the NWT mixed economy, emphasizing the settler colonial material impetus for the high rates of violence against Indigenous women. That said, Indigenous women across Canada are working to socially reproduce their communities through various relationships with capital, outside of the context of a mixed economy. What is generalizable here is the racism that makes embodied violence against Indigenous women permissible, and the work of socially reproducing a minority community a site of de/colonizing struggle.

In interviews and talking circles, women described their experiences of violence, from sexual harassment at the mine site (capitalist production); to intimate partner violence in relationships where women feel trapped due to, among other things, financial inequalities (social reproduction); to malnutrition as a result of the degradation of country food (subsistence production). Violence to the land – in the form of extraction – was experienced as structural violence insofar as it was violent against subsistence production and Indigenous place-based systems of meaning. At the same time, violence against the land was found to be an embodied violence – a violence that has been experienced as malnutrition, illness, and even death. For many research participants, the impact of the diamond mines is not felt as an isolated, or unique, phenomenon. Rather, it has been experienced throughout the twentieth century *relationally*, as a series of violent interventions into Indigenous women's lives and labour. This continuity was expressed, for example, when Jenna ruminated on whether the trauma experienced in her community was the result of residential schools, the diamond mines, or both (Interview 109). This serves to remind us of the mixed economy's resilience in the face of multiple ruptures. While I have characterized the diamond-mining regime as violent, I have been careful to avoid characterizing that regime as *only* violent. Indeed, the diamond mines have provided increased material security for some as well as a range of experiences for the people in the region of study. Because I built this research through existing community partnerships, often forged through shared interests in gender justice and challenging violence against Indigenous women, it can be assumed that the participants' contributions to this book are skewed toward a concern for the gendered impacts of the mines, and particularly its violences. With respect to Indigenous women's experiences of the diamond mines, the research participants expressed the ways in which they navigated the violence of the diamond mines, and resisted that violence, yet chose to engage with the regime when they felt it was necessary or beneficial for themselves, their families, and/or their community.

Diverse experiences with the diamond mines and the prevalence of acts of creation and resistance do not, however, diminish the violent tendencies of the diamond-mining regime. Informed by anti-racist theories that position colonialism and racism as violence (Fanon 1963; Goldberg 1993), and by feminist theories linking capitalism with gender violence (Agathangelou 2004; Federici 2004; Hill-Collins 2006; Kuokkanen 2011), I concur that there is a general violence, as well as a gendered and racialized violence, to capitalism. Building on this assumption, I have examined the relationship between gender violence and FIFO extraction in the context of the mixed economy. The finding that there is a violence to FIFO echoes political-economic and social research that has tracked the psychosocial impacts of FIFO labour on mine workers, their families, and mining communities (for example, Government of Australia 2014). However, my approach has uniquely attended to the ways in which this violence plays out through the relations of capitalist production, social reproduction, and subsistence production. I have argued that there is a racialized gender violence to the separation of capitalist production (in the form of extraction) from social reproduction. The FIFO mines are a spatial expression of capital's gendered tendency to separate capitalist production from social reproduction, a separation operationalized through the feminization and naturalization of social reproduction. At the same time, resource extraction persists in tension with subsistence economies, and in undermining subsistence relations, it enacts a racialized violence. Clearly, this racialized violence does not operate independently of the gendered violence that occurs when capitalist production is separated from social reproduction. Rather, these are two dialectical aspects of a structural violence that has shaped Indigenous women's relationships to the diamond mines. Indeed, the separation of capitalist production from social reproduction is made possible by the racialized marginalization of subsistence, as a form of labour that denies the boundary between production and social reproduction.

The violence of FIFO extraction is also an embodied violence. The FIFO model has the potential to create sites of hypermasculinity: spaces outside of community-based gender norms with the capacity to perpetuate their own site-specific culture. Research participants noted shifts, for example, in their partners' views about women after they had spent time at the diamond mines. Community workers discussed the

problem of reaching diamond mine workers with anti-violence education initiatives.[1] These specific examples echo scholarly work characterizing mine sites as hypermasculine spaces (Scott 2007), isolated as they
are, dominated by men, and with histories associating masculinity with
extraction. This dynamic is especially pernicious in the context of the
recent history of settler colonial rupture in Indigenous social relations
in the Yellowknife region. Thus, while there is a generalizability to the
embodied violence of FIFO, there is also a settler colonial specificity; in
the region of study, the relationship between embodied violence and
mining has been structured through the intense and ongoing violence
of the settler colonial intervention in the North. The destruction of land
and disruption of animals by the diamond mines (most notably, the caribou herds), and the resulting impact on subsistence production and,
thus, the health and well-being of northern Indigenous people, is perhaps the most visceral example of the links between the structural and
embodied impacts of ongoing settler colonialism.

III. Diamonds Are Not Forever

Violence was far from the only experience that research participants
discussed in relation to the diamond mines. The narratives they shared
were profound stories of strength, community, and resistance. Thus,
settler colonial violence is best understood in terms of processes and
embodiments of rupture rather than as annexation or destruction. The
temporal imperatives of the diamond-mining regime – its temporariness, and its tempo – clashed with the day-to-day and intergenerational
social reproduction of Indigenous place-based ontologies and subsistence. And the diamond mines are, indeed, that: temporary, a rupture
to be accounted for in the longer history of subsistence-oriented social
relations around Yellowknife. As noted by John B. Zoe (in Gibson 2008),
eras in Tłı̨chǫ history can be mapped through shifts in relationships
(initiations of new relationships, negotiations of difference, and resolution); whether and how the diamond mines will be remembered in
terms of this history is yet to be seen. What is certain is that Indigenous
women in the Yellowknife region are looking ahead to the post-diamond
mixed economy.

Indeed, one of the most common emergent themes – particularly in
talking circles – was the question, "What will happen after the diamond
mines close?" A singular feature of the FIFO diamond-mining regime
is that its presence has thus far been as brief as it is deep; Snap Lake
Diamond Mine, the most recent diamond mine to open, closed unex-
pectedly in 2015. Dominion has announced that it will close in 2025,

while Ekati is still attempting to prolong its mine life. It is predicted that Gatcho Kué, opened in 2015, will have a twelve-year mine life. So a number of research participants spoke to the post-diamond question: What will the regional political economy look like after diamonds? Will there be another extractive boom? Will the region sink into a post-extractive depression? Or will communities build a future beyond extraction, one with a greater emphasis on subsistence and community well-being, a differently mixed economy?

The restructuring of the diamond mines can be grafted onto multiple, intersecting narratives of time and place: the FIFO tempo that has shaped the lives of so many residents of the area around Yellowknife; the abrupt ups and downs of the brief life of the FIFO diamond-mining regime; the modern settler history of Yellowknife, and its continuity as a mining town; and the place-based narratives of Dene history, as a shifting of relationships, not an end to them. In locating the rupture of the diamond-mining regime in a larger context, my aim is not to dull the edge of its racialized gendered violence or its restructuring of social relations more generally. Rather, the relational narratives of multiple forms of labour shared by Dene, Métis, and Inuit women present experiences that challenge and expand upon Western capitalist typologies of production and social reproduction and that elevate subsistence as a contemporary reality and future possibility. Indigenous modes of life threaten the totality of capital and the sovereignty of the settler state by demonstrating that another way is possible. My aim has been to honour these forms of labour that have represented – throughout the development of the gold mines and the diamond mines, and the Canadian state's various approaches to northern "development" – decolonizing relations tied to the same place across changing times. For while diamonds have been imbued with all kinds of powerful characteristics, from a marker of love to the driver of the northern economy, in the end, they are, as Annie reminded us in a talking circle, "just rocks." "Diamonds, they're just rocks," she told us. "They're just rocks. Think of the water. Water is life. You look at the plants, you don't water them, you die ... A rock is not going to feed you. The wildlife is. The water is" (Talking Circle One).

Appendix

List of Interviews, Talking Circles, and Focus Groups

* Interview 100's = women who self-identified as having a direct relationship to the diamond mines; Interview 200's = community workers; Interview 300's = territorial government employees; Interview 400's = diamond industry management.

- Interview 101: 5 June 2014. Yellowknife.
- Interview 102: 7 June 2014. Yellowknife.
- Interview 103: 11 June 2014. Yellowknife.
- Interview 104: 11 June 2014. Yellowknife.
- Interview 105: 20 June 2014. Yellowknife.
- Interview 106: 25 June 2014. Behchoko.
- Interview 107: 25 June 2014. Behchoko.
- Interview 108: 25 June 2014. Behchoko.
- Interview 109: 25 June 2014. Behchoko.
- Interview 110: 27 June 2014. Yellowknife.
- Interview 111: 9 July 2014. Yellowknife.
- Interview 112: 25 June 2014. Behchoko.
- Interview 113: 11 July 2014. Yellowknife.
- Interview 114: 14 July 2014. Yellowknife.
- Interview 115: 16 July 2014. Yellowknife.
- Interview 116: 17 July 2014. Yellowknife.
- Interview 117: 21 July 2014. Yellwoknife.
- Interview 118: 22 July 2014. Yellowknife.
- Interview 119: 1 August 2014. Yellowknife.
- Interview 120: 12 August 2014. Yellowknife.
- Interview 201: 9 June 2014. Yellowknife.
- Interview 202: 9 June 2014. Yellowknife.

- Interview 203: 18 June 2014. Yellowknife.
- Interview 204: 20 June 2014. Yellowknife.
- Interview 205: 25 June 2014. Behchoko.
- Interview 206: 23 July 2014. Yellowknife.
- Interview 207: 11 August 2014. Yellowknife.
- Interview 301: 10 July 2014. Yellowknife.
- Interview 302: 16 July 2014. Yellowknife.
- Interview 303: 8 August 2014. Yellowknife.
- Interview 304: 13 August 2014. Yellowknife.
- Interview 401: 24 July 2014. Yellowknife.
- Interview 402: 28 July 2014. Yellowknife.
- Talking Circle 1: 3 July 2014. Yellowknife.
- Talking Circle 2: 5 August 2014. Yellowknife.
- Focus Group 1: 9 July 2014. Yellowknife.

Notes

1 Introduction

1 Material quoted in this paragraph is from "The Cremation of Sam McGee" by Robert Service.

2 See, for example, former prime minister Stephen Harper's claim that "the true North is our destiny – for our explorers, for our entrepreneurs, for our artists. To not embrace the promise of the true North, now, at the dawn of its ascendency, would be to turn our backs on what it is to be Canadian" (Government of Canada 2009: 3). Harper's assertion obscures the activist role the Canadian state and private interests have played in northern development in Canada's modern history, in particular since the Second World War (Watkins 1977; DiFrancesco 2000; Abele 2009a).

3 The density of reality TV based in Yellowknife certainly demonstrates a persistent fascination with the North as an imagined frontier. Yellowknife, a town of 20,000, is home to three network reality TV series: *Ice Road Truckers*, *Ice Pilots*, and *Ice Lake Rebels*, all of which depict a harsh, mostly empty landscape that requires a rugged, intrepid spirit.

4 In using the term "anachronistic space," I am referring to Anne McClintock's critique of imperialism and colonialism (1995). She writes: "Since indigenous peoples are not supposed to be spatially there – for the lands are 'empty' – they are symbolically displaced onto what I call anachronistic space ... According to this trope, colonized people – like women and the working class in the metropolis – do not inhabit history proper but exist in a permanently anterior time within the geographic space of the modern empire as anachronistic humans, atavistic, irrational, bereft of human agency – the living embodiment of the archaic 'primitive'" (30).

5 The concept "place-based" draws upon Glen Coulthard's (2010) discussion of the unique role that place – in contrast to the Western linear focus on time – plays in Indigenous belief systems and social relations.

6   Just over 40,000 people live in the NWT, around half of whom are Indigenous. Most of NWT's Indigenous people live in thirty-two smaller communities (GNWT 2015), while most of its non-Indigenous people live in Yellowknife. Unlike in Yukon, the NWT's urban population is spread out among a few towns (Coates and Powell 1989). Hay River, Fort Smith, and Inuvik all have populations between 2,500 and 4,000 and are much larger than most Indigenous communities. They also have significant non-Indigenous populations. Most Indigenous communities have fewer than 1,000 inhabitants, with the smallest, Kakisa, having around fifty-five residents. Note also that some writings count as many as sixty NWT communities, depending on how communities are defined, "demonstrating that Northern Indigenous ways of moving and settling simply do not fit neatly into Western data collection" (Hall 2013: 391).

7   Dettah, N'dilo, and Yellowknife are Yellowknives Dene communities now part of the Akaitcho Dene First Nation, while Behchokǫ̀ is the site of government for the Tłı̨chǫ Nation.

8   Kulchyski is referring, in particular, to N'dilo's proximity to Old Town, a neighbourhood that, as the name suggests, is the oldest (non-Indigenous) part of town. One can still find these earliest shacks around Old Town, beside some of the most expensive homes in town.

9   Hiring through the temporary foreign worker program has been restricted by the federal government due to high rates of unemployment in the territory; however, the city of Yellowknife applied for and was granted an exemption (GNWT 2015b).

10 According to GNWT data, around 25 per cent of Yellowknife residents identify as Indigenous. Also, 18 per cent of Yellowknife adults speak an Indigenous language well enough to carry on a conversation (GNWT 2015a), compared to 38 per cent of the population as a whole. The Indigenous languages include Inuktitut, Inuvialuktun, Inuinnaqtun, Dogrib, Cree, Chipewyan, North Slavey, South Slavey, and Gwich'n. However, the full-time Indigenous population in Yellowknife at any one time is bolstered by the many transients who arrive from their home communities in the NWT for any length of time for work, health and social service appointments, education, or visits with family and friends. This transience, a particular feature of Yellowknife, is important for understanding the relationship between Yellowknife and the smaller Indigenous communities in the North.

11 The questions that guided the first community focus group were written in collaboration with the Native Women's Association. They were: (1) How have your relationships and your family been impacted by the diamond mines? (2) What challenges do the diamonds bring to community well-being? (3) How do diamond mines impact children?

(4) How do diamond mines impact women and men differently? And (5) How has your life been impacted by the diamond mines? The questions for the second focus group were only slightly revised: (1) How has your life been changed by the diamond mines? (2) How do the diamond mines impact women and men differently? (3) What challenges [do the diamond mines] bring to community well-being? How have your relationships and your family been impacted by the diamond mines?

12 In implementing fieldwork, I adjusted this target. Because of the participation of government workers in focus groups, and my access to government policy documents, I determined that I had collected the data required for analysis.

2 An Expanded Approach to Production

1 Marx continues, "It is no longer sufficient, therefore, for him [the worker] simply to produce. He must produce surplus-value. The only worker who is productive is one who produces surplus-value for the capitalist, or in other words, contributes towards the self-valorization of capital" (1976: 644). Because this definition is in keeping with a general theoretical consensus on "capitalist production" in critical political economic literature (Fine and Filho 2010), I do not elaborate further on this concept.

2 For the purposes of the following discussion, it is useful to explain Marx's definition of capitalist value, as it is distinct from common usage. For Marx, capitalist value is measured not by its social necessity but by "the labour time socially necessary to produce it, including both direct (living) and indirect (dead, or past) labour inputs" (Fine and Filho 2010: 18). As such, new value is created under capitalism through the exploitation of labour; that is to say, through the various means of extracting more labour than that which has been compensated.

3 This is not to say that Marx ignores colonization as an aspect of capitalism; in his chapters on primitive accumulation, Marx demonstrates that the annexation of new spaces of capital accumulation is part of the capitalist mode of production (1976).

4 This, I would argue, was the primary weakness of the domestic labour debates, the major FPE contribution to the analysis of reproduction in the 1970s and 1980s. The "debate" of the domestic labour debates refers to a series of theoretical arguments over whether work in the home was productive labour, in the Marxian sense, and generally focused on work in one's own home (which implicitly obscured the labour of women – primarily women of colour – working for pay in other people's homes). Interestingly, while Seccombe's later work is little concerned with

questions of productivity or non-productivity, he was a major contributor to these earlier debates (1973, 1975).

5 This will be discussed in detail in chapter 3; however, it is worth noting that, according to the GNWT (2015), around 30 per cent of NWT households accessed more than half their food through traditional activities (this is known as "country food").

6 Examples of capitalist Indigenous labour in the NWT include Indigenous subcontractors, which sometimes operate along principles distinct from those of more typical Western mining operations. The diamond-mining companies also employ small, Indigenous-owned and -operated businesses as a strategy for avoiding labour in the extractive industry and for fostering cultural sustenance.

7 See, for example, Milloy (1999), Kelm (1998), and Fournier and Crey (1997). These are important decolonizing interventions, for Indigenous ill health (much like violence) is often read either as an individual medical problem or as a determinant of dysfunction.

8 For an articulation of the North as a colony, see, for example, Kenneth Coates (1985). This is not to say that Coates is unaware of Indigenous issues; this is not the case. Rather, I am referring specifically to the way in which he characterizes northern autonomy.

9 A concern aptly raised by Abele and Stasiulis (1989).

10 In speaking to the multiple ways in which people build their lives in the North, it is worth also noting the myriad local institutions that contribute to social reproduction at the community level. Recent social research in the NWT has come to use the term "social economy" when analysing the mixed economy (Kuokkanen 2011), the myriad forms and levels of state, Indigenous, and community northern governance (Abele 2009a, 2009b), and local for-profit and not-for-profit organizations oriented toward community well-being. It is beyond the scope of this project to engage with this rich body of scholarship; however, it is a reminder to attend to the complex ways in which non-Indigenous and Indigenous people are working outside of the imperatives of capital in building northern lives.

3 Wıìlıìdeh's Mixed Economy

1 By "newly racialized," I am not suggesting that settler colonial processes of racialization began in this region in the twentieth century – rather, that the socio-economic processes of the time led to new developments in the ways in which Indigenous social relations in the region of study were racialized.

2 Indeed, Coulthard writes, 'Indigenous struggles against capitalist imperialism are best understood as struggles oriented around the question

of land – struggles not only for land, but also deeply informed by what the land as a mode of reciprocal relationship (which is itself informed by place-based practices and associated forms of knowledge) ought to teach us about living our lives in relation to one another and our surroundings in a respectful, non-dominating and nonexploitative way" (2014: 60).

3 Caribou are very important for the Dene, materially and symbolically. This long-standing relationship to the caribou is important to note, given the impacts the diamond mines have had on caribou migration patterns.

4 It is the anachronistic account of culture that is problematic here, as certainly this labour *is* imbued with cultural meaning. However, to use its cultural meaning to assign this labour a space outside of relations of production is to deny its role in the maintenance and social reproduction of the mixed economy.

5 As noted in chapter 2, this is not to say that these labours are not interdependent or that the boundary between the two is impermeable; only that there is a real distinction in their relationship to the valuation of capital.

6 Gail Kellough (1980) provides a useful overview of analyses of Indigenous incorporation or non-incorporation into the Canadian capitalist economy, arguing that Indigenous people were not left out of economic development, but rather were incorporated as an underclass. Wotherspoon and Satzewich (2000) also review this literature: in particular, they critique models of "internal colonialism" for inaccurately omitting Indigenous agency from analysis and argue for a political economic model with "an emphasis on the changing material circumstances which shape and are shaped by aboriginal life experience."

7 Indeed, both Kulchyski (2005) and Coulthard (2014) note that, as settlers tended to be relied upon for labour, northern Indigenous people experienced the impact of capitalism as dispossession rather than proletarianization. The specific approach to Indigenous labour taken up by the gold mines is discussed in more detail in chapter 4.

8 The Canadian state's strategic interest in the North refers to a new concern for sovereignty brought about principally by a concern regarding American military presence north of the 60th parallel during and following the Second World War (Coates 1985; Abele 2009a), as well as to an intensified interest in northern resource extraction (DiFrancesco 2000).

9 It is beyond the scope of this chapter to discuss it in detail, but the fur trade had a major impact on the lives of Indigenous people in the NWT. For an analysis of this, see Asch (1977), Coates (1985), and Blondin (1997). Regarding the impact on Indigenous people, Coates (1985) writes that "they gained access to the technology and material devices of a different age and civilization [sic], but they also faced the devastation

caused by imported disease, a depletion of game through over-hunting and considerable economic and social change" (30). Settler fur trades established ten camps, or forts, along the Mackenzie River (Gibson 2008: 54); Indigenous peoples engaged in the fur trade shifted their annual travel patterns to participate in these camps, the locations of which have shaped colonial settlements in the NWT to this day. Given the subject of this book, it is of note that Ron Bourgeault (1983) argues that participation in the fur trade undermined egalitarian relationships between women and men in northern Indigenous communities because men were seen as responsible for the fur as a tradable commodity and thus "assumed the role of the head of the household," while women were targeted by settlers as a means of accessing Indigenous communities. Bourgeault's claims are provocative insofar as they offer an assessment of the way in which gendered relationships shifted through early contact between northern Indigenous peoples and settlers; however, his insistence on the degradation of gender equality and communal structures of reproduction and production is challenged by many accounts of Indigenous social relations of that time (see Asch 1977), including the Yellowknives Dene First Nations Elders Advisory Council, who point to significant changes in their land use patterns as a result of the fur trade, but not to changes in their gendered structures of social reproduction or production.

10 A third site, Negus Mine, operated from 1939 to 1952 (Silke 2009).

11 Jack Lambert moved to Yellowknife from Edmonton, where he was making $35 per month. At Con Mine, Lambert started at $125 per month (Silke 2009: 14).

12 It is worth noting that, while I characterize the work as stable and well-paid, this does not negate the difficult and sometimes dangerous nature of extractive labour. Indeed, while not all extractive jobs or mine sites are the same, the bodily toll taken for the pursuit of extractive capital is significant. This has been discussed, in particular, in relation to work at Fort McMurray and in relation to questions of violence and the embodied impacts of colonial restructuring of reproduction and production.

13 The United Steelworkers of America took over in 1968.

14 I focus on union organizing, in large part, to offer a comparative between the gold mining extractive regime and the diamond mining extractive regime, which do not share the strong links between workers and union as a result of both the spatial separation inherent to FIFO and neoliberal tendencies toward subcontracting.

15 Today, the Con Mine head frame, or shaft, towers above Yellowknife, It is surrounded by controversy over whether it should be preserved using city tax dollars. Those in favour argue that the head frame is an irreplaceable

piece of the city's history – an interesting debate given who was and was not given access to the head frame during its years of use.

16 This approach to economic development is noteworthy, in particular as it emphasizes the political and social nature of the shift toward economic development-as-capitalist-employment that occurs through the diamond-mining regime. Given government's activist role in the development of the regional political economy around Yellowknife, there is nothing "natural" about the new recruitment of Indigenous people into the extractive industry.

17 This quotation from Marie Adele Sangris is part of Elders Tape 1968/1972 transcribed by Lena Drygeese in 1993. It was included in a presentation by the Yellowknives Dene bringing to light the devastation caused by the gold mines on their lands, communities, and ways of life.

18 The arsenic used to extract gold from the Giant Mine remains in the land around Yellowknife to this day; specifically, there are 237,000 tonnes stored in permafrost chambers under Yellowknife. As noted by local environmental and social justice activists (O'Reilly 2012) and national and international media, this is enough arsenic to kill every person on the earth several times over (CBC 2014; *Huffington Post* 2014).

19 See Dickerson (1992) for a comprehensive history and policy analysis of NWT federal administration and points of devolution.

20 See Wotherspoon and Satzewich (2000) for an analysis of the federal bureaucracy as it relates to Indigenous administration.

21 The constitutional basis for territorial self-government was established by the NWT Act, written in 1875 and amended in 1905 to reflect new territorial boundaries (Dickerson 1992; Abele 2009a).

22 In April 2014, jurisdiction over subsurface rights was granted to the NWT. This has important implications for the management and distribution of revenue from the diamond mines, which is discussed in chapter 5.

23 While the commentary on the skunk is amusing, to be sure, I include it for its representation of an imagined "old Yellowknife" reacting to the new, professionalized workforce arriving in town. The domesticated animal – in contrast to the northern, undomesticated raven – demonstrates a tension that persists in Yellowknife to this day.

24 Bean writes that "in retrospect, the move from camps to centralized settlements had great significance for the native peoples, not the least of which was the establishment of the dominance of the government and the reciprocal dependency of native peoples" (1977: 130).

25 The move to government-subsidized housing would come to have important implications for the impact of diamond mining on conditions for social reproduction. Because rent is based on a sliding scale, a significant number of participants discussed situations wherein a

household would go from paying $27 per month to $1,500 per month for the same place.

26 The administration of these schools was moved from the federal government to the territorial government in 1967.

27 Della's name has not been changed as she asked that her own name be attached to this story.

28 Della first told this story in 2014 when we were preparing a presentation on residential schools. When I asked Della for permission to share her story in this venue, she agreed and generously wrote it out in her own words.

29 MacGregor (2010) is analysing education in Nunavut. The context is different in Nunavut, as education is governed by the territory, which is Inuit. However, the tension MacGregor outlines exists in the NWT as well, though it plays out through distinct multi-level governmental structures.

30 It is noteworthy that 22,000 is the number of all Indigenous people in the Territory, not of Indigenous children. As such, the percentage of Indigenous children who are receiving services would be significantly higher.

4 The Global Political Economy of Canadian Diamonds

1 In his text, *Fire on Earth*, McCarthy discusses the early history of diamond mining.

2 Appadurai goes on to write: "Better still, since most luxury goods are used (though in special ways and at special cost), it might make more sense to regard luxury as a special 'register' of consumption (by analogy of the linguistic model) than to regard as a special class of thing. The signs of this register, in relation to commodities are some or all of the following attributes: (1) restriction, either by price or by law, to elites; (2) complexity of acquisition, which may or may not be a function of real "scarcity"; (3) semiotic virtuosity, that is, the capacity to signal fairly complex social messages (as do pepper incuisine, silk in dress, jewels in adornment, and relics in workshop); (4) specialized knowledge as a prerequisite for their 'appropriate' consumption, that is, regulation by fashion; and (5) a high degree of linkage of their consumption to body, person, and personality" (1986: 38).

3 The tar sands are also defended by their related jobs, an issue I will discuss in the following chapter.

4 Here, it is useful to think of the difference between "necessary" (recalling that necessities are, themselves, socially constructed) and luxury goods as a set of signifiers falling along a spectrum, rather than a dichotomy. Certainly, daily goods have the potential to capture aspirations as well, as, for example, Biro (2019) has demonstrated with water consumption.

5 This is to say that the "legality" or "illegality" of mining projects is a contested and shifting terrain, wherein international politics (for example, Foreign Direct Investment agreements) can make international mining projects violating domestic law legal, and the surrounding politics of small-scale mining, either in accessible deposits or in previously abandoned mines, can contribute to its perception as, alternatively, an example of "responsible" development (i.e., community-led artisanal mining) or as illegal.

6 Ian Smillie's many works on this subject provide an excellent insider account of these processes.

7 Concerns about uneven development between core and periphery continue to colour development language in the North; however, there is a discrepancy between those who see the North, its governments, and all northern residents as part of the periphery (the implication here is that resource rents must stay in the North – the approach of the current territorial government), and those who focus on the uneven exploitation of northern Indigenous populations through an analysis of settler colonialism.

8 At the time of writing, the territorial government continues to initiate incentives to entice industries auxiliary to the diamond industry (CBC 2018).

9 See Butler (2015) for a critical analysis of Canadian mining and "international development."

5 The NWT Diamond-Mining Regime

1 I discuss the deaths of the nine workers here to stress the severity of the violence that marked the end of the gold-mining regime. I do so with respect to all the people in Yellowknife impacted by this violence, recognizing that the strike and the deaths of the workers continue to be a source of pain for many. The purpose here is not to pass judgment on the different factions of this time.

2 As so often happens with settler tales of "exploration" and "discovery," the role these land experts played in finding the diamond deposit has been written out of contemporary diamond folklore.

3 Recall the discussion in chapter 3 of the multiple concepts of "colonialism" as they relate to the North.

4 All GNWT images and figures have been reproduced here with the generous permission of the GNWT.

5 This is the value of exports from the NWT in 2013 (the last year the federal government collected 100% of resource royalties), 99.8% of which was accrued through the sale of diamonds.

6 These are the Inuvialuit Regional Corporation, Sahtu Secretariat Inc., Tłı̨chǫ Government, Gwich'in Tribal Council, Northwest Territories Metis Nation, Salt River First Nation, Deninu K'ue First Nation, Acho Dene Koe First Nation, and Katl'odeeche First Nation.

7 While, at the time of writing, the act has not been finalized, the draft has been criticized by Yellowknife MLA Kevin O'Reilly for its limited and unclear regulation of the mining industry (Brockman 2019). Indeed, the Act will likely perpetuate the free-entry approach common to extraction in Canadian legislation and regulation, which has been intensified in the NWT.

8 The length of a "mine life" is not fixed and can shift based on local and global political economic developments – most significantly, market value of the mineral. For example, the 2008–9 financial crisis led to temporary camp closures at both Ekati and Diavik and concerns that the two mines would have significantly shorter lives than expected. Currently, however, developments of new sites at Ekati are expected to extend the mine another ten years.

9 Indigenous activists responded to the 1969 White Paper with the Red Paper, and the proposal was buried under widespread criticism and dedicated activism.

10 As I argue elsewhere (Hall 2015), the federal government's interest in securing certainty in land tenure was expressed contemporaneously in Harper's pursuance of conversion of Indigenous land tenure into fee-simple land. Under liberal private property ownership arrangements, Indigenous land can more effectively be exploited for capitalist accumulation.

11 As Irlbacher-Fox (2009) notes, "although land claim negotiations and agreements are closely related to those of self-government, land claim and self-government rights and authorities as understood by the Canadian state are distinct and until the 1995 Inherent Right Policy were dealt with separately" (2).

12 As noted by some interview participants, the definition of "northerner" is a site of contention, particularly for those who view the North as an internal colony, as discussed in chapter 1 and articulated by Coates (1985). The concern is that loose definitions of northerner enable both transient northern residents – who come to the North for a job in the mines and leave quickly – and people who work at the mines but move to southern Canada to benefit from northern preferential hires.

13 In her account of the Łutselk'e community experience of IBA negotiations, Bielawski (2003) illustrates the ways in which the confidentiality of agreements enables diamond companies to isolate communities in negotiation, thereby presenting the "choice" to consent to diamond

mining as a choice between mining occurring and a community receiving benefits or mining occurring with no benefits going to the community. However, as noted by Gibson (2008: 134) and some interview participants, arguably, the confidentiality of IBAs protects Indigenous communities from government clawbacks of transfer payments based on their resource revenue sources.

14 The role of IBAs in community consultation, the substance of their negotiation and implementation, and the extent to which they actually represent new, more "responsible" development is a rich space of on-the-ground policy work and academic inquiry (see, for example, O'Faircheallaigh 2006, 2007; Laforce et al. 2009; Caine and Krogman 2010).

15 See Cornell and Kalt (1998); Anderson and colleagues (2006); and Altamarino-Jiménez (2004) for a critique of a neoliberal market approach to Indigenous development.

16 New mining operations currently under development are discussing the option of offering Indigenous communities company equity rather than cash payments or socio-economic commitments to the community (Interview 402). This is a step further in the direction of tying Indigenous community material interests to the success of extractive projects.

17 Hache, an interview participant, is quoted using her real name, as she indicated she did not wish to be quoted anonymously.

6 Time, Place, and the Diamond-Mining Regime

1 While it is beyond the scope of this analysis to discuss, certainly capitalism is not uniform and innovations in both fixed capital and modes of exploiting variable capital can result in increased surplus-value.

2 This is only partly true. Recall that this book approaches resource extraction as a social phenomenon, acknowledging that minerals are mined not just because they are there but because they are in spaces deemed extractable (namely, not the centres, but rather the peripheries of empire). In the Canadian context, Bay Street is not going to be dug up anytime soon, no matter what lies beneath.

3 Certainly, I am not diminishing the long history of economic and social depression in mining towns post-closure, nor the tendency of mining companies to skip town in these moments. Rather, I am noting that FIFO obscures and erodes the relationship between company and town. This tendency is mitigated, to some degree, in the NWT context because of the norms for Indigenous consultation discussed in the previous chapter.

4 Nanisivik was a zinc/lead mine with a company town in Nunavut (what was the NWT at the time of the mine opening). It operated from 1976 to 2002.

5 There is a short-term winter ice road connecting Diavik and Ekati to Yellowknife (Gibson 2008: 17), used for the transportation of heavy equipment and resources to the mines. However, people are transported to the mines almost exclusively by air.

6 Mary Lou Cherwaty did not wish to be quoted anonymously.

7 For example, Jenny, who worked for a sub-contractor, described the disparity between contract work and working for the mine:

> J: Basically all the people who were doing contract work...we were kind of separate from actual Diavik.
> R: That's interesting. So your trailer was separate. Did you eat separately?
> J: We eat separate from the Diavik staff. We ate with the contractor staff ... It was actually a separate cafeteria ... The only reason why I knew that is that I actually knew people at the time who were working for Diavik so I would occasionally try and sneak in. Yeah, because the contractor side was a little bit rougher and there were virtually zero women. It will honestly be a memory I carry with me till the day I die. A sea of men and there was only myself and one other woman who worked on my cross-shift. (Interview 115)

8 This picture, of course, only covers capitalist production. Those who are characterized as "not working" because of their exclusive engagement in social reproduction or non-capitalist subsistence work are discussed in chapter 8.

9 This comparison is not a critique of the administration of Aurora College; like many organizations in the NWT, they have responded to the funding parameters delineated by the federal and territorial government.

10 Indeed, a community-built Facebook group, titled "NWT cost of living is out of control!," chronicles exorbitant food prices across the territory (https://www.facebook.com/groups/264091951563/).

11 This characterization does not refer to the medium-sized towns of the territory: Hay River, Fort Smith, Inuvik, or Norman Wells.

12 Indeed, mining company policies around criminal record (i.e., that a person, for the most part, cannot have a record to be employed) have been a recognized barrier to employment in the context of the NWT, where Indigenous people are vastly overrepresented in the criminal justice system (as in the rest of Canada). This speaks to the implications of the colonial political economy of incarceration.

13 See Paula Butler's (2015) work for a discussion of the Canadian state's implementation of this strategy in their "gender equality" initiatives attached to a Canadian gold mine in Burkina Faso.

14 In chapters 7 and 8, it will become clear that these responsibilities for social reproduction extend beyond the household and involve

community-level social reproduction oriented toward subsistence – labour that is a resistance to the totalizing impulses of capital.

15 And indeed, Wright (1999) makes the link between the "disposability" of "untrainable" women workers in the *maquiladora* sector and the abhorrent rates of murder against women in northern Mexico.

16 Embodied violence refers to "deliberate behavior in which one person chooses to dominate, control or harm another" (Weaver et al. 2007: 5), including, but not limited to psychological, physical, and sexual abuse and harassment.

7 Social Reproduction and the Diamond-Mining Regime

1 While I discuss embodied violence in this chapter, and the context of this violence and insights that emerge from research participants, I omit details of research participants' actual experiences of physical violence for ethical reasons: that is, I do not believe *specifics* of personal experience of physical or sexual violence will significantly strengthen the analysis here and I believe there is the potential to retraumatize the research participants, or other readers of this project.

2 Certainly, it is striking that no participants described the benefits of diamond-mine employment without qualifying those benefits or also discussing the negative impacts or dangers. However, it is worth recalling here the selection bias of the interviewees: research participants volunteered to be interviewed or to participate in a focus group or talking circle, and as such, it is possible that those who came forward were people with issues around the diamond mines that they wished to voice.

3 In pointing to this tendency, I am drawing on place-specific observations from interviews, as well as feminist literature on heightened gender inequality in times of crisis (see, for example, Rai 2002; Kuokkonen 2008)

4 This is Marx's labour theory of value (1976): that is, that labour-power, and specifically, unpaid labour-power, is the source of new value in capitalism.

5 Indeed, both Federici (2004) and Maria Mies (1986) theorize the medieval witch hunts as primarily attacks on women's reproductive autonomy by a patriarchal and newly capitalist state. Mies writes that "the witch hunt had not only the direct disciplinary effect of controlling women's sexual and reproductive behaviour, but also the effect of establishing the superiority of male productivity over female productivity. These two processes are closely connected" (70).

6 This finding emerged in cases where a male partner was working in the diamond mines and had a female partner taking on the primary caregiver role. All research participants interviewed were in male–female

domestic partnerships, though this should not be taken as representative of coupling trends in the NWT. It is important to note that, when the roles were reversed – that is, when women worked at the diamond mines while men stayed at home – the trend did not remain the same (though almost all research participants were women and the dynamics in households wherein women worked at the mines and men stayed at home, or that involved same-sex relationships, were not specifically investigated). Indeed, research participants who had both worked at the diamond mines and acted as primary caregiver while their partner worked at the mine spoke of a continuity of a responsibility for care, which demonstrates that the feminized responsibility for care is both an ideological process and a material one.

7 The most substantive body of literature on the links between FIFO and gender violence relates to Australia, a major FIFO country. Concerns that FIFO employment leads to higher rates of violence (Sharma and Rees 2007) are often linked to concerns about substance use, depression, and strains on families and romantic relationships (Greer and Stokes 2011).

8 Certainly, state retrenchment of social services has increased social reproduction responsibilities for northern women (both Indigenous and not), as I allude to here; but it is important not to conflate the experiences of retrenchment by non-racialized and racialized women, given that Indigenous women continue to feel a heavy state presence in their lives, often in the form of surveillance over their modes of social reproduction.

9 Concerns about newly high rates of cancer in communities, and their possible links to the diamond mines, were expressed by a number of informants. Indigenous communities have endured the devastating long-term health impacts of extractive projects all over Canada. Ward Churchill (2011), for example, has documented the high incidence of cancer and birth abnormalities in Indigenous communities as a result of uranium mining (2001); Brittany Luby (2015) has looked to the gendered health impacts (particularly on prenatal health) of hydroelectric damming on Anishnabe communities; and a number of community activists in the NWT are struggling to document the impact of the arsenic deposits left by gold mines in Yellowknife (O'Reilly 2012).

8 Diamonds, Subsistence, and Resistance

1 Here, Usher and colleagues (2003) are referring to state research, not scholarly research.

2 It is notable that, since these data do not capture the consumption of country food in households where its percentage is less than 50 per cent, a great deal of subsistence consumption is missed here.

3 That is to say that neither labour nor subjectivities fall into neat categories in the northern mixed economy. For example, settlers have a long history of engaging with their Indigenous friends and neighbours and learning subsistence activities.

4 It is of note that the GNWT has not published these data since 2015. The *Communities and Diamonds* report has replaced this set of stats monitoring the traditional economy with data on fur sales (included as what they call "one window" into the traditional economy), and now, a community wellness index that includes engagement with traditional activities in its definition of "wellness." This move away from attention toward traditional activities, limited from the outset, is troubling.

5 Here they are discussing not just the Sahtu but the entire NWT.

6 While the insight into the potential relations between subsistence production and capitalist production is important, it is worth noting that the Sahtu region described here has a relationship to wage labour that is quite distinct from the political economy of the region around Yellowknife, the political and business centre of the territory.

7 Mies and Bennholdt-Thomsen make this argument in a general way, but draw their generalizations largely from European peasant economies.

8 That is, that the exploitation of wage labour is internal to the dynamics of capitalist production and thus becomes business-as-usual rather than an exploitation that requires consistent violence and coercion.

9 That is, through a social capital lens, "resilience" can be used as a rhetorical device to download supportive and health responsibilities onto a community. However, "resilience" is also used by northern and Indigenous scholars to point to the existing strengths of Indigenous communities (Irlbacher-Fox 2009).

10 It speaks, I would argue, at least in part, to the intensity of the disciplining, blaming, and surveilling of Indigenous bodies and health that Sarah felt the need to explain that her daughter's diabetes was not the result of obesity (which is often read as the consequence of poor individual choices, particularly for Indigenous bodies).

11 For example, when asked about the impacts of the diamond mines, Adam, a local non-Indigenous environmental researcher said, "So, diamond mines aren't actually so bad. Because it's really compact. Yeah, you're not affecting a big area. The biggest impacts are fuel use ... The diamond mines are actually better than the gold mines. It's hard rock, so there's a bit of acid drainage, but it doesn't have the impacts from cyanide and some of the more [dangerous chemicals], like arsenic" (Interview 203).

12 The remediation of Giant Mine is an ongoing site of contention; while mining and government officials insist that Yellowknife and the surrounding communities are safe from the arsenic deposits, Indigenous

and environmental groups continue to struggle for better protections and oversight over the project.

9 Conclusion

1 This concern was also raised in regard to anti-violence legislation. Specifically, community workers noted incidents where Emergency Protection Orders (EPOs) were issued after men used violence against their partners, but the RCMP were not able to serve EPOs because the men were at camp (Focus Group).

References

Abele, F. 2006. "Indigenous People in the Cities of Northern Canada: The Importance of the Rural Economic Base." In *New Actors in Northern Federations: Cities, Mergers, and Aboriginal Governance in Russia and Canada*, ed. Peter H. Solomon, Jr. Toronto: Centre for European, Russian and Eurasian Studies, University of Toronto.

– 2009a. "North Development: Past, Present and Future." In *Northern Exposure: Peoples, Powers and Prospects in Canada's North*, edited by Frances Abele, Thomas J. Courchene, F. Leslie Seidle, and France St.-Hilaire, 19–65. Montreal: Institute for Research on Public Policy.

– 2009b. "The State and the Northern Social Economy: Research Prospects." *Northern Review* 30 (Spring 2009): 37–58.

– 2015. "State Institutions and the Social Economy in Northern Canada." In *Northern Communities Working Together: The Social Economy of Canada's North*, edited by Chris Southcott, 74–96. Toronto: University of Toronto Press.

Abele, F., and D. Stasiulis. 1989. "Canada as a 'White Settler Colony': What about Natives and Immigrants?" in *The New Canadian Political Economy*, edited by W. Clement and G. Williams, 240–77. Montreal and Kingston: McGill–Queen's University Press.

Adams, H. 1995. *A Tortured People: The Politics of Colonization*. Penticton: Theytus Books.

Agathangelou, A. 2004. *The Global Political Economy of Sex: Desire, Violence, and Insecurity in Mediterranean Nation-States*. New York: Palgrave Macmillan.

Adams, H. 1995. *A Tortured People: The Politics of Colonization*. Penticton: Theytus Books.

Alfred, T. 2005. *Was se: Indigenous Pathways of Action and Freedom*. Toronto: University of Toronto Press.

Alfred, T. 2010. "What Is Radical Imagination? Indigenous Struggles in Canada." *Affinities: A Journal of Radical Theory, Culture, and Action* 4(2): 5–8.

Alfred, T., and J. Corntassel. 2005. "Being Indigenous: Resurgences against Contemporary Capitalism." *Government and Opposition* 40(4): 597–614. https://doi.org/10.1111/j.1477-7053.2005.00166.x.

Altamirano-Jiménez, I. 2004. "North American First Peoples: Slipping Up into Market Citizenship?" *Citizenship Studies* 8(4): 349–65. https://doi.org /10.1080/1362102052000316963.

Amnesty International. 2009. *Canada: Follow-Up to the Concluding Observations of the United Nations Committee on the Elimination of Discrimination against Women*. London.

– 2014. *Violence against Indigenous Women and Girls in Canada: A Summary of Amnesty International's Concerns and Calls to Actions*. IWFA Submission. February 2014. London.

– 2016. *The Point of No Return: The Human Rights of Indigenous Peoples in Canada Threatened by the Cite C Dam*. London.

Anderson, K. 2003. "Vital Signs: Reading Colonialism in Contemporary Adolescent Family Planning." In *Strong Women Stories: Native Vision and Community Survival*, edited by K. Anderson and B. Lawrence, 173–91. Toronto: Sumach Press.

Anderson, K., and B. Lawrence, eds. 2003. *Strong Women Stories: Native Vision and Community Survival*. Toronto: Sumach Press.

Anderson, R., L.P. Dana, and T.E. Dana. 2006. "Indigenous Land Rights, Entrepreneurship, and Economic Development in Canada: 'Opting-in' to the Global Economy." *Journal of World Business* 41: 45–55. https://doi.org /10.1016/j.jwb.2005.10.005.

Appadurai, A., ed. 1988. *The Social Life of Things: Commodities in Cultural Perspective*. Cambridge: Cambridge University Press.

Asch, M. 1977. "The Dene Economy." In *Dene Nation: The Colony Within*, edited by Mel Watkins. Toronto: University of Toronto Press.

Bakker, I. 2003. "Neo-Liberal Governance and the Reprivatization of Social Reproduction: Social Provisioning and Shifting Gender Orders." In *Power, Production, and Social Reproduction*, edited by I. Bakker and Gill, 55–82. London: Palgrave Macmillan UK.

Banaji, J. 1977. "Modes of Production in a Materialist Conception of History." *Capital and Class* 1(3): 1–44. https://doi.org/10.1177/030981687700300102.

Bannerji, H. 2005. "Building from Marx: Reflections on Class and Race." *Social Justice* 32(4): 144–60.

Barry, P.S. 1992. "The Canol Project, 1942–45." *Arctic* 45(4): 401–3. https://doi .org/10.14430/arctic1420.

Baskin, C. 2003. "From Victims to Leaders: Activism against Violence Towards Women." *Strong Women Stories: Native Vision and Community Survival*, edited by K. Anderson and P. Lawrence, 213–27. Toronto: Sumach Press.

Bean, W. 1977. In *Dene Nation: The Colony Within*, edited by Mel Watkins. Toronto: University of Toronto Press.

Beers, R. 2018, 10 October. "Halted Social Work Program at Aurora College Should Be Expanded, Report Suggests." *CBC News*.

Bell, L. 2013. *Diamonds as Development: Suffering for Opportunity in the Canadian North*. PhD diss., University of Toronto.

Bellamy Foster, J., and B. Clark. 2004. "Ecological Imperialism: The Curse of Capitalism." *Socialist Register* 40: 186–201.

Bezanson, K., and M. Luxton. 2006."Introduction: Social Reproduction and Feminist Political Economy." In *Social Reproduction: Feminist Political Economy Challenges Neo-Liberalism*, edited by K. Bezanson and M. Luxton, 3–10. Montreal and Kingston: McGill–Queen's University Press.

Bhattacharya, T., ed. 2017. *Social Reproduction Theory: Remapping Class, Recentering Oppression*. London: Pluto Press.

Bielawski, E. 2003. *Rogue Diamonds: Northern Riches on Dene Land*. Seattle: University of Washington Press.

Bieri, F. 2010. "The Roles of NGOs in the Kimberley Process." *Globality Studies Journal* 20: 1–13.

Biro, A. 2019. "Reading a Water Menu: Bottled Water and the Cultivation of Taste." *Journal of Consumer Culture* 19(2): 231–51. https://doi.org/10.1177 /1469540517717779.

Blackstock, C. 2016. "The Complainant: The Canadian Human Rights Case on First Nations Child Welfare." *McGill Law Journal* 62(2): 285–328. https://doi .org/10.7202/1040049ar.

– 1997. *Yamoria the Lawmaker: Stories of the Dene*, vol. 1. Edmonton: NeWest Press.

Blondin, G. 1997. *Yamoria the Lawmaker: Stories of the Dene*. Edmonton: NeWest Press.

Bone, R., and M.B. Green. 2003. "The Northern Native Labour Force: A Disadvantaged Work Force." In *Canada's Changing Workforce*, edited by W. Wonders, 380–7. Montreal and Kingston: McGill-Queen's University Press.

Bourgeault, R. 1983. "The Indian, the Metis, and the Fur Trade: Class, Sexism, and Racism in the Transition from 'Communism' to Capitalism." *Studies in Political Economy* 12 (Fall): 45–80. https://doi.org/10.1080/19187033.1983.1 1675649.

Brave Heart, Maria Yellow Horse. 2003. "TheHistorical Trauma Response among Natives and Its Relationship with Substance Abuse: A Lakota Illustration." *Journal of Psychoactive Drugs* 35(1): 7–13. https://doi.org /10.1080/02791072.2003.10399988. Medline: 12733753.

Brockman, A. 2019. "Mixed Reaction as Review of NWT's Mineral Resources Act Begins." *CBC North News*, 13 February.

Butler, P. 2015. *Colonial Extractions: Race and Canadian Mining in Contemporary Africa*. Toronto: University of Toronto Press.

Caine, K., and N. Krogman. 2010. "Powerful or Just Plain Power-Full? A Power Analysis of Impact and Benefit Agreements in Canada's North."

Organization and Environment 23(1): 76–98. https://doi.org/10.1177/1086026609358969.

Cameron, E., and T. Levitan. 2014. "Impact and Benefit Agreements and the Neoliberalization of Resource Governance and Indigenous–State Relations in Northern Canada. *Studies in Political Economy* 93(1): 25–52. https://doi.org/10.1080/19187033.2014.11674963.

Carey Consulting and Evaluation 2011. *Northern Women in Mining, Oil, and Gas Project: Summative Project Evaluation for the Reporting Period April 2007–February 2010. Yellowknife, Northwest Territories.* Commissioned by the Status of Women Council of the NWT. Iqaluit.

Carrier, J.G. 2010. "Protecting the Environment the Natural Way: Ethical Consumption and Commodity Fetishism." *Antipode* 4(3): 672–89. https://doi.org/10.1111/j.1467-8330.2010.00768.x.

Carrón-Prieto, M., M. Thomson, and M. Macdonald. 2007. "No More Killings! Women Respond to Femicides in Central America." *Gender and Development* 15(1): 25–40. https://doi.org/10.1080/13552070601178849.

CBC (Canadian Broadcasting Corporation). 2014, 28 November. "High Arsenic Levels Found in Some Yellowknife-Area Lakes."

– 2018, 6 December. "N.W.T. Gov't Loosening Rules in Hopes of Wooing Northern Diamond Manufacturers."

Chamberlain, M.E. 2014. *The Scramble for Africa.* New York and London: Routledge.

CEDAW (Convention on the Elimination of all Forms of Discrimination Against Women). 2010, 9 February. *Interim Report in Follow-Up to the Review of Canada's Sixth and Seventh Reports.* CEDAW/C.CAN/CO/7/Add.1. New York.

Christensen, J.. 2014. "Our Home, Our Way of Life: Spiritual Homelessness and the Sociocultural Dimensions of Indigenous Homeless in the Northwest Territories (NWT), Canada." *Social and Cultural Geography* 14(7): 804–28. https://doi.org/10.1080/14649365.2013.822089.

– 2017. *No Home in a Homeland: Indigenous Peoples and Homelessness in the Canadian North.* Vancouver: UBC Press.

Churchill, W. 2001 *Struggle for the Land: Native North American Resistance to Genocide, Ecocide, and Colonization* [2001]. San Francisco: City Lights.

Coates, K. 1985. *Canada's Colonies: A History of the Yukon and Northwest Territories.* Toronto: James Lorimer.

Coates, K., and J. Powell. 1989. *The Modern North: People, Politics, and the Rejection of Colonialism.* Toronto: James Lorimer.

Colletti, L. 1972. *From Rousseau to Lenin.* New York: Monthly Review Press.

Collymore, S. 1980. *Native Labour in the Northern Mining Industry.* http://www.miningnorth.com/resources.

Cornell, S., and J.P. Kalt. 1998. "Sovereignty and Nation-Building: The Development Challenge in Indian Country Today." *American Indian Culture*

and Research Journal 22(3): 187–214. https://doi.org/10.17953/aicr.22.3
.lv45536553vn7j78.

Cortright, D., G.A. Lopez, R.W. Conroy, J. Dashti-Gibson, J. Wagler, D.M.
Malone, and L. Axworthy. 2000. *The Sanctions Decade: Assessing UN
Strategies in the 1990s*. Boulder: Lynne Rienner.

Coulthard, G. 2010. "Place against Empire: Understanding Indigenous Anti-
Colonialism." *Affinities: A Journal of Radical Theory, Culture, and Action* 4(2):
79–83.

– 2014. *Red Skin, White Masks: Rejecting the Colonial Politics of Recognition*.
Minneapolis: University of Minnesota Press.

Coulthard, G., and L.B. Simpson. 2016. "Grounded Normativity/Place-Based
Solidarity." *American Quarterly* 68(2): 249–55. https://doi.org/10.1353/aq
.2016.0038.

Davis, A. 1981. *Women, Race, and Class*. New York: Random House.

De Beers Group. 2018. *The Diamond Insight Report 2018*. debeersgroup.com.

Dickerson, M. 1992. *Whose North: Political Change, Political Development, and
Self-Government in the Northwest Territories*. Vancouver: UBC Press.

DiFrancesco, Richard. 2000. "A Diamond in the Rough?: An Examination of
the Issues Surrounding the Development of the Northwest Territories,"
Canadian Geographer 44(2): 114–34. https://doi.org/10.1111/j.1541-0064.2000.
tb00697.x.

Dombrowski, K., B. Khan, E. Channell, J. Moses, K. McLean, and E. Misshula.
2013. "Kinship, Family, and Exchange in a Labrador Inuit Community."
Arctic Anthropology 50(1): 89–104. https://doi.org/10.3368/aa.50.1.89

Edwards, T. 2019, 21 January. "YKU." Edgenorth.ca.

Ellem, B. 2013. "Peak Union Campaigning: Fighting for Rights at Work in
Australia." *British Journal of Industrial Relations* 51(2): 264–87. https://doi
.org/10.1111/j.1467-8543.2011.00878.x.

Epstein, E. 1982, February. "'Have you ever tried to sell a diamond?'" *The
Atlantic*.

Falvo, N. 2011. *Homelessness in Yellowknife: An Emerging Social Challenge*.
Toronto: Canadian Homelessness Research Network Press.

Fanning, I. 2018. *Ininí Ádisókán (Man Stories): Masculinities among the
Mámíwinini (Algonquin People)*. PhD diss., Queen's University.

Fanon, F. 1963. *The Wretched of the Earth*. New York: Grove Press.

Federici, S. 2004. *Caliban and the Witch: Women, the Body, and Primitive
Accumulation*. New York: Autonomedia.

Ferguson, S. 2008. "Canadian Contributions to Social Reproduction Feminism,
Race, and Embodied Labor." *Race, Gender, and Class* 15(1–2): 42–57.

Ferguson, S., and D. McNally. 2013. "Foreword." In L. Vogel, *Marxism and the
Oppression of Women: A Unitary Theory*. Chicago: Haymarket Books.

Fine, B., and S. Filho, A. 2010. *Marx's Capital*. London: Pluto Press.

Fitzpatrick, P. 2007. "A New Staples Industry? Complexity, Governance, and Canada's Diamond Mines." *Policy and Society* 26(1): 93–112. https://doi .org/10.1016/S1449-4035(07)70102-9.

Fournier, S., and E. Crey. 1997. *Stolen from Our Embrace: The Abduction of First Nations Children and the Restoration of Aboriginal Communities.* Vancouver: Douglas and McIntyre.

Galeano, E. 1997. *The Open Veins of Latin America: Five Centuries of the Pillage of a Continent* [1976]. New York: NYU Press.

Gibson, V. 2008. *Negotiated Spaces: Work, Home, and Relationships in the Dene Diamond Economy.* PhD diss., University of British Columbia.

Gibson MacDonald, G., J.B. Zoe, and T. Satterfield. 2014. "Reciprocity in the Canadian Dene Diamond Mining Economy." In *Natural Resource Extraction and Indigenous Livelihoods: Development Challenges in an Era of Globalisation,* edited by E. Gilberthorpe and G.M. Hilson. Surrey: Ashgate.

Gilbert, C. 2015. "Update: Snap Lake Diamond Snaps Shut, 434 Laid Off." *Northern Journal.* 7 December. http://norj.ca/2015/12/de-beers -halts-mining-at-snap-lake.

Global Witness. 1998. *A Rough Trade: The Role of Companies and Governments in the Angolan Conflict.* London.

Goldberg, D.T. 1993. *Racist Culture.* Cambridge: Blackwell.

Gordon, T., and J.R. Webber. 2008. "Imperialism and Resistance: Canadian Mining Companies in Latin America." *Third World Quarterly* 29(1): 63–87. https://doi.org/10.1080/01436590701726509.

Government of Australia. 2014. *Shedding a Light on FIFO Mental Health: A Discussion Paper.* Report no. 4. Canberra: Education and Health Standing Committee.

Government of Canada. 2009. *Canada's Northern Strategy: Our North, Our Heritage, Our Future.* Ottawa: Indian Affairs and Northern Development [*sic*] and Federal Interlocutor for Metis and Non-Status Indians.

– 2014. *Canadian Trade and Investment Activity. Northwest Territories' Merchandise Trade with the World.* Ottawa.

– 2018a. "Reducing the Number of Indigenous Children in Care." *First Nations Child and Family Services.* Ottawa. https://www.sac-isc.gc.ca/eng/154118735 2297/1541187392851.

– 2018b, 7 September. *Guide to the Labour Force Survey.* Ottawa: Statistics Canada.

Government of the Northwest Territories (GNWT). 1993. *Dene Kede – Education: A Dene Perspective.* Yellowknife: Education, Culture, and Employment.

– 1996. *Inuuqatigiit: The Curriculum from the Inuit Perspective.* Yellowknife: Education, Culture, and Employment.

– 2012. *The Residential School System in Canada: Understanding the Past – Seeking Reconciliation – Building Hope for Tomorrow.* Yellowknife: Education, Culture and Employment.

– 2013a. *Communities and Diamonds*. Yellowknife: Departments of Industry, Tourism, and Investment; Education, Culture, and Employment; Finance; Health, and Social Services; Justice; Statistics; and NWT Housing Corporation.
– 2013b. *Dene Kede – Education: A Dene Perspective*. Yellowknife: Department of Education, Culture, and Employment.
– 2014a. *Communities and Diamonds*. Yellowknife: Departments of Industry, Tourism and Investment; Education, Culture and Employment; Finance; Health and Social Services; Justice; Statistics; NWT Housing Corporation
– 2014b. *Community Data*. Yellowknife: Bureau of Statistics.
– 2014c. *Education Renewal and Innovation Framework: Directions for Change*. Yellowknife: Department of Education, Culture, and Employment.
– 2015a. *Community Data*. Yellowknife: Bureau of Statistics.
– 2015b. "Minister Announces Exemption to Temporary Foreign Worker Program for Employers in Yellowknife." 27 April. Yellowknife: Education, Culture, and Employment.
– 2016. *Community Data*. Yellowknife: Bureau of Statistics.
– 2018a. *Communities and Diamonds Socio-Economic Agreements Annual Report 2017*. Yellowknife: Departments of Industry, Tourism and Investment; Education, Culture and Employment; Finance; Health and Social Services; Justice; Statistics; NWT Housing Corporation.
– 2018b. *Community Data*. Yellowknife: Bureau of Statistics.
– 2019a. *2018 Socio-Economic Agreement Report For Diamond Mines Operating in the Northwest Territories*. 28 May. Yellowknife: NWT Legislature.
– 2019b. *Bill 34: Mineral Resources Act* (DRAFT). First reading. 11 February. Yellowknife: NWT Legislature.
– 2019c. *Diamond Policy Framework*. Yellowknife: Industry, Tourism and Investment.
Government of the Northwest Territories Legacy and Hope Foundation. 2013. *The Residential School System in Canada: Understanding the Past, Seeking Reconciliation, Building Hope for Tomorrow*. Yellowknife.
Government of the Northwest Territories and Nunavut Chamber of Mines. 2014. *NWT Mineral Development Strategy*. Yellowknife: Department of Industry, Tourism and Investment.
Grant, J.A. 2012. "The Kimberley Process at Ten: Reflections on a Decade of Efforts to End the Trade in Conflict Diamonds." In *The Global Diamond Industry*, 119–42. London: Palgrave Macmillan
Grant, J.A., and I. Taylor. 2004. "Global Governance and Conflict Diamonds: The Kimberley Process and the Quest for Clean Gems." *The Round Table* 93(375): 385–401. https://doi.org/10.1080/0035853042000249979.
Green, D. 2015. Personal correspondence.
Greer, L., and K. Stokes. 2011. "Divorce and Separation in the Australian Mining Sector: Is It What We Expect." *Report for the Minerals Futures*

Collaboration Cluster in CSIRO Minerals Down Under National Research Flagship Regions in Transition: Social Impacts of Mining.

Hall, S. 1986. "Gramsci's Relevance for the Study of Race and Ethnicity." *Journal of Communication Inquiry* 10(2): 5–27. https://doi.org/10.1177/019685998601000202.

– 2013. "Diamond Mining in Canada's Northwest Territories: A Colonial Continuity." *Antipode: A Radical Journal of Geography* 45(2): 376–93. https://doi.org/10.1111/j.1467-8330.2012.01012.x.

– 2015. "Divide and Conquer: Privatizing Indigenous Land Ownership as Capital Accumulation." *Studies in Political Economy* 96: 23–45. https://doi.org/10.1080/19187033.2015.11674936.

Hann, M., D. Walsh, and B. Neil. 2014. "At the Crossroads: Geography, Gender, and Occupational Sector in Employment-Related Geographical Mobility." *Canadian Studies in Population* 41(3–4): 6021. https://doi.org/10.25336/P6G60D.

Harnum, B., J. Hanlon, T. Lim, J. Modeste, D. Simmons, and A. Spring. 2014. *Best of Both Worlds: Sahtú Goné T'áadets'enı̨ o̧ – Depending on the Land in the Sahtú Region*. Tulita: Pembina Institute.

Hartwick, E.R. 2000. "Towards a Geographical Politics of Consumption." *Environment and planning A* 32(7): 1177–92. https://doi.org/10.1068/a3256.

Hawkeye Robertson, L. 2006. "The Residential School Experience: Syndrome or Historical Trauma." *Pimatisiwin* 4: 2–28.

Hill-Collins, P. 2006. "New Commodities, New Consumers: Selling Blackness in a Global Marketplace." *Ethnicities.* 6(3): 297–317. https://doi.org/10.1177/1468796806068322.

Hoogeveen, D. 2016. "Subsurface Property, Free-Entry Mineral Staking, and Settler Colonialism in Canada." *Antipode* 47(1): 1. https://doi.org/10.1111/anti.12095.

hooks, b. 1984. *Feminist Theory: From Margin to Center*. Boston: South End Press.

Huffington Post. 2014. "Yellowknife Mine Sits Atop a Lethal Store of Arsenic. 17 April.

Hunn, E.S. 1999. "The Value of Subsistence for the Future of the World." In *Ethnoecology: Situated Knowledge/Located Lives*, ed. Virginia D. Nazarea, 23–36. Tuscon: University of Arizona Press. https://doi.org/10.2307/j.ctv1gwqrkg.6.

Hurcomb, F. 2012. *Old Town: A Photographic Journey through Yellowknife's Defining Neighbourhood*. Yellowknife: Outcrop.

Innis, H. 1927. *The Fur Trade in Canada: An Introduction to Canadian Economic History*. Toronto: University of Toronto Press.

ILO (International Labour Organization). 2018. *World Employment Social Outlook: Trends 2018*. Geneva.

Irlbacher-Fox, S. 2009. *Finding Dahshaa: Self-Government, Social Suffering, and Aboriginal Policy in Canada*, Vancouver: UBC Press.

Irlbacher-Fox, S., and S.J. Mills. 2009. "Living Up to the Spirit of Modern Treaties? Implementation and Institutional Development." In *Northern Exposures: Peoples, Powers, and Prospects in Canada's North*, edited by F. Abele, T.J. Courchene, F.L. Seidle, and F. St.-Hilaire, 19–65. Montreal: Institute for Research on Public Policy.

Jenson, J. 1986. "Gender and Reproduction or Babies and the State." *Studies in Political Economy* 20: 9–46. https://doi.org/10.1080/19187033.1986.11675588.

Jessop, B. 1982. *The Capitalist State: Marxist Theories and Methods*. Oxford: Martin Robertson.

Jiwani, Y. 2006. *Discourses of Denial: Mediations of Race, Gender, and Violence*. Vancouver: UBC Press.

Kakfwi, S., and B. Overvold. 1977. "The Schools." In *Dene Nation: The Colony Within*, edited by Mel Watkins, 142–8. Toronto: University of Toronto Press.

Karl, T.L. 1999. "The Perils of the Petro-State: Reflections on the Paradox of Plenty." *Journal of International Affairs* 53(1): 31–48.

Kellough, G. 1980. "From Colonialism to Economic Imperialism: The Experience of the Canadian Indian." In *Structured Inequality in Canada*, edited by J. Harp and J. Hofley, 343–77. Toronto: Prentice-Hall.

Kelm, M.E. 1999. *Colonizing Bodies: Aboriginal Health and Healing in British Columbia, 1900–1950*. Vancouver: UBC Press.

King, H. 2013. "New Treaties, Same Old Dispossession: A Critical Assessment of Land and Resource Management Regimes in the North." In *Canada: The State of the Federation 2013: Aboriginal Multilevel Governance*, edited by Martin Papillon and André Juneau, 83–96. Montreal and Kingston: McGill-Queen's University Press.

Kino-nda-niimi Collective. 2014. *The Winter We Danced: Voices from the Past, the Future, and the Idle No More Movement*. Winnipeg: ARP Books.

Kitzinger, J. 1994. "The Methodology of Focus Groups: The Importance of Interaction between Research Participants." *Sociology of Health and Illness* 16(2): 103–21. https://doi.org/10.1111/1467-9566.ep11347023.

Kohl, B., and L. Farthing. 2012. "Material Constraints to Popular Imaginaries: The Extractive Economy and Resource Nationalism in Bolivia." *Political Geography* 31(4): 225–35. https://doi.org/10.1016/j.polgeo.2012.03.002.

Kulchyski, P. 2005. *Like the Sound of a Drum: Aboriginal Cultural Politics in Denendeh and Nunavut*. Winnipeg: University of Manitoba Press.

Kuokkanen, R. 2008. "Globalization as Racialized, Sexualized Violence: The Case of Indigenous Women." *International Feminist Journal of Politics* 10(2): 216–33. https://doi.org/10.1080/14616740801957554.

– 2011. "Indigenous Economies, Theories of Subsistence, and Women." *American Indian Quarterly* 35(2): 215–40. https://doi.org/10.5250/amerindiquar.35.2.0215.

Laforce, M., U. Lapointe, and V. Lebuis. 2009. "Mining Sector Regulation in Quebec and Canada: Is a Redefinition of Asymmetrical Regulations Possible?" *Studies in Political Economy* 84: 47–78. https://doi.org/10.1080/19187033.2009.11675046.

Lawrence, B. 2003. "Gender, Race, and the Regulation of Native Identity in Canada and the United States: An Overview." *Hypatia* 18(2): 3–31. https://doi.org/10.1353/hyp.2003.0031.

Le Billon, P. 2001. "The Political Ecology of War: Natural Resources and Armed Conflict." *Political Geography* 20: 561–84. https://doi.org/10.1016/S0962-6298(01)00015-4.

– 2006. "Fatal Transactions: Conflict Diamonds and the (Anti)terrorist Consumer." *Antipode* 38(4): 778–801. https://doi.org/10.1111/j.1467-8330.2006.00476.x.

Legat, A. 2012. *Walking the Land, Feeding the Fire: Knowledge and Stewardship among the Tlicho Dene.* Tucson: University of Arizona Press.

Little, L. 2007. *Securing Our Place in Northern Societ: Women, Global Industries, and the Power of Stories.* MA final project, submitted to Integrated Studies, Athabasca University.

– 2010. *Newcomers' Initiative: Final Report.* Submitted to the NWT Literacy Council, Yellowknife.

Luby, B. 2015. "From Milk-Medicine to Public (Re)Education Programs: An Examination of Anishinabek Mothers' Responses to Hydroelectric Flooding in the Treaty #3 District, 1900–1975." *Canadian Bulletin of Medical History* 32(5): 363–89. https://doi.org/10.3138/cbmh.32.2.363. Medline: 28155379.

Luxton, M. 2006. "Feminist Political Economy in Canada and the Politics of Social Reproduction." In *Social Reproduction: Feminist Political Economy Challenges Neoliberalism,* edited by K. Bezanson and M. Luxton. Montreal and Kingston: McGill-Queen's University Press.

MacGregor, H. 2010. *Inuit Education and Schools in the Eastern Arctic.* Vancouver: UBC Press.

Maracle, L. 1988. *I Am Woman.* Vancouver: Write-On Press.

Markey, S. 2010. "Fly-In, Fly-Out Resource Development: A New Regionalist Perspective on the Next Rural Economy." In *The Next Rural Economies: Constructing Rural Place in Global Economies,* edited by G. Halseth, S. Patrick Markey, and D. Bruce, 238–50. CABI Publishing.

Markey, S., K. Storey, and K. Heisler. 2011. "Fly In/Fly Out Resource Development: Implications for Community and Regional Development." In *Demography at the Edge: Remote Human Populations in Developed Nations,* edited by D. Carson, R.O. Rasmussen, P. Ensign, L. Husley, and A. Taylor, 213–36. New York and London: Routledge.

Marx, K. 1971. *The Grundrisse.* New York: Harper Torchbooks.

– 1976. *Capital,* vol. 1 [1867]. London: Penguin Books.

McArthur, D. 2009. "The Changing Architecture of Governance in the Yukon and the Northwest Territories." *Northern Exposure: Peoples, Powers, and Prospects in Canada's North*, edited by F. Abele, T.J. Courchene, F.L. Seidle, and F. St.-Hilaire, 19–65. Montreal: The Institute for Research on Public Policy.

McCarthy, J. 1942. *Fire in the Earth: The Story of the Diamond*. Whitefish: Literary Licensing.

McClintock, A. 1995. *Imperial Leather: Race, Gender, and Sexuality in the Colonial Contest*. New York and London: Routledge.

McDonald, P., R. Mayes, and B. Pini. 2012. "Mining Work, Family, and Community: A Spatially-Oriented Approach to the Impact of the Ravensthorpe Nickel Mine Closure in Remote Australia." *Journal of Industrial Relations* 54(1): 22–40. https://doi.org/10.1177/0022185611432382.

Mies, M. 1986. *Patriarchy and Accumulation on a World Scale: Women in the International Division of Labour*, 3rd ed. London: Zed Books, 2014.

Mies, M., and Veronika Bennholdt-Thomsen. 1999. *The Subsistence Perspective: Beyond the Globalised Economy*. Melbourne: Spinifex Press.

Miller, J. 1996. *Shingwauk's Vision: A History of Native Residential Schools*. Toronto: University of Toronto Press.

Milloy, J.S. 1999. *A National Crime: The Canadian Government and the Residential School System, 1879 to 1986*, vol. 11. Winnipeg: University of Manitoba Press.

MMIWG (National Inquiry into Missing and Murdered Indigenous Women). 2019. *Reclaiming Power and Place: The Final Report of the National Inquiry into Missing and Murdered Indigenous Women and Girls*.

Moore, J.W. 2003. "The Modern World-Systems Environmental History? Ecology and the Rise of Capitalism." *Theory and Society* 32(3): 307–77. https://doi.org/10.1023/A:1024404620759.

Nahanni, P. 1977. "The Mapping Project." In *Dene Nation: The Colony Within*, edited by Watkins. Toronto: University of Toronto Press.

– 1992. *Dene Women in the Traditional and Modern Northern Economy in Denendeh, Northwest Territories Canada*. MA thesis, Department of Geography, McGill University.

NRCAN. 2019. https://www.nrcan.gc.ca/our-natural-resources/minerals-mining/mining-resources/importing-and-exporting-rough-diamonds-the-kimberley-process/8222.

NWT and Nunavut Chamber of Mines. 2008. *Mining and Exploration: Northwest Territories 2008 Overview*. Yellowknife.

– 2017. *Measuring Success 1996–2016: Diamond Mines Deliver Big Benefits to the Northwest Territories*. Yellowknife.

– 2018. *25 Years of Diamonds: Discovery, Development, Economy, Innovation, Legacy, Future*. Yellowknife. www.miningnorth.com.

NWT Housing Commission. 2015. *Northwest Territories Housing Corporation Annual Report*. Yellowknife.

NWT Union of Northern Workers. 2014. *Internal Reports*. Provided in fieldwork.

O'Faircheallaigh, C. 2006. "Mining Agreements and Aboriginal Economic Development in Australia and Canada." *Journal of Aboriginal Economic Development* 51(1): 74–91.

– 2007. "Environmental Agreements, EIA Follow-Up, and Aboriginal Participation in Environmental Management: The Canadian Experience." *Environmental Impact Assessment Review* 27(4): 319–42. https://doi.org /10.1016/j.eiar.2006.12.002.

Office of the Auditor General of Canada. 2014. *Report of the Auditor General of Canada to the Northwest Territories Legislative Assembly, Child and Family Services*. Yellowknife.

O'Reilly, K. 2012. "Giant Mine, Giant Legacy." *Northern Public Affairs* 1(2): 50–3.

Outcrop. 2000. *Yellowknife Tales: Sixty Years of Stories from Yellowknife*. Yellowknife.

Parlee, B. 2015. "The Social Economy and Resource Development in Northern Canada." In *Northern Communities Working Together: The Social Economy of Northern Canada*, edited by C. Southcott, 52–73. Toronto: University of Toronto Press.

Peck, J. 2001. *Workfare States*. New York: Guilford.

– 2013. "Excavating the Pilbara: A Polanyian Exploration." *Geographical Research* 51(3): 227–42. https://doi.org/10.1111/1745-5871.12027.

Peoples Jeweller. 2019. https://www.peoplesjewellers.com/collections/arctic -brilliance-canadian-diamonds/c/3251922.

Pini, B., and R. Mayes. 2014. "Performing Rural Masculinities: A Case Study of Diggers and Dealers." In *Masculinities and Place*, edited by Andrew Gorman-Murray and Peter Hopkins, 431. New York and London: Routledge, 2018.

Price, R. 1974. *Yellowknife*. Greencastle: Buckingham Books.

Rae, K.J. 1976. *The Political Economy of Northern Development* Ottawa: Science Council of Canada.

Rai, S. 2002. *Gender and the Political Economy of Development*. Cambridge: Polity Press.

Rainnie, A., S. Fitzgerald, B. Ellem, and C. Goods. 2014. "FIFO and Global Production Networks: Exploring the Issues." *Australian Bulletin of Labour* 40(2): 98.

Razack, S., ed. 2002. *Race, Space, and the Law: Unmapping a White Settler Society*. Toronto: Between the Lines.

Rodriguez, S. 2012. *The Femicide Machine*. Los Angeles: Semiotext(e).

Sabin, J. 2014. "Contested Colonialism: Responsible Government and Political Development in Yukon." *Canadian Journal of Political Science* 47(2), 375–96. https://doi.org/10.1017/S0008423914000419.

Saku, J.C. 1999. "Aboriginal Census Data in Canada: A Research Note." *Canadian Journal of Native Studies*, 19(2): 365–79.

Sangris, M.A. 1968/1972. *Elders Tape*. Transcribed by Lena Drygeese. Yellowknive Dene.

Santarossa, B. 2004. *Diamonds: Adding Lustre to the Canadian Economy*. Ottawa: Statistics Canada.

Schlosser, K. 2013. "History, Scale, and the Political Ecology of Ethical Diamonds in Kugluktuk, Nunavut." *Journal of Political Ecology* (20)1: 53–69. https://doi.org/10.2458/v20i1.21746.

Scott, R. 2007. "Dependent Masculinity and Political Culture in Pro-Mountaintop Removal Discourse: or, How I Learned to Stop Worrying and Love the Dragline." *Feminist Studies* 33(3): 484–509.

Seccombe, W. 1973. "The Housewife and Her Labour under Capitalism." *New Left Review* 1(83): 3–24.

– 1975. "Domestic Labour – Reply to Critics." *New Left Review* 1(94): 84–96.

– 1986. "Patriarchy Stabilized: The Construction of the Male Breadwinner Wage Norm in Nineteenth-Century Britain." *Social History* 11.(1): 53–76. https://doi.org/10.1080/03071028608567640.

– 1992. *A Millennium of Family Change: Feudalism to Capitalism in Northwestern Europe*. New York and London: Verso.

Selleck, L., and F. Thompson. 1997. *Dying for Gold: The True Story of the Giant Mine Murderer*. Toronto: HarperCollins Canada.

Sharma, S., and S. Rees. 2007. "Consideration of the Determinants of Women's Mental Health in Remote Australian Mining Towns." *Australian Journal of Rural Health*, 15(1): 1–7.

Shewell, H. 2004. *"Enough to Keep Them Alive": Indian Welfare in Canada, 1873–1965*. Toronto: University of Toronto Press, 2004.

Silke, R. 2009. "A History of Aboriginal Participation in Mining." In *Up Here Business*. Yellowknife: Outcrop.

Simpson, A. 2016. "The State Is a Man: Theresa Spence, Loretta Saunders, and the Gender of Settler Sovereignty." Theory and Event 19(4). Retrieved from https://www-proquest-com.proxy.queensu.ca/scholarly-journals /state-is-man-theresa-spence-loretta-saunders/docview/1866315122 /se-2?accountid=6180.

Simpson, L. 2014. "Not Murdered, Not Missing." *Nations Rising*. Posted on 5 March.

Sisters in Spirit. 2010. *What Their Stories Tell Us: Research Findings from the Sisters in Spirit Initiative*. Akwesasne Reserve No. 15: Native Women's Association of Canada.

Smillie, I. 2010. *Blood on the Stone: Greed, Corruption, and War in the Global Diamond Trade*. London: Anthem Press.

Smillie, I., L. Gberie, and R. Hazleton. 2000. *The Heart of the Matter: Sierra Leone, Diamonds, and Human Security: Complete Report*. Collingdale: Diane.

Smith, A. 2005. *Conquest: Sexual Violence and American Indian Genocide*. Cambridge, MA: South End Press.

Soederberg, S. 2014. *Debtfare States and the Poverty Industry: Money, Discipline, and the Surplus Population*. New York and London: Routledge.

Southcott, C., ed. 2015. *Northern Communities Working Together: The Social Economy of Canada's North*. Toronto: University of Toronto Press.

Speakman, M. 2014. Public talk. Yellowknife: Native Women's Association of the NWT.

Statistics Canada. 2013. *Family Violence in Canada: Statistics Profile*. Ottawa: Canadian Centre for Justice Statistics.

Starblanket, G. 2017. "Being Indigenous Feminists: Resurgences against Contemporary Patriarchy." In *Making Space for Indigenous Feminism*, edited by J. Green, 21–62. Halifax: Fernwood.

Storey, K. 2010. "Fly-In/Fly-Out: Implications for Community Sustainability." *Sustainability* 2(5): 1161–81. https://doi.org/10.3390/su2051161.

Stueck, W. 2002. "Canadian Quality Adds Sparkle to Diamonds." *Globe and Mail Report on Business*. 23 November.

Thobani, S. 2007. *Exalted Subjects: Studies in the Making of Race and Nation in Canada*. Toronto: University of Toronto Press.

Tienhaara, K. 2011. "Regulatory Chill and the Threat of Arbitration: A View from Political Science." In *Evolution in Investment Treaty Law and Arbitration*, edited by C. Brown and K. Miles. Cambridge: Cambridge University Press.

Trochu, M. 2017. "Federal Gov't Gives $7.4M to Mine Training Society to Train Indigenous People." *CBC News North*. 13 October.

True, J. 2012. *The Political Economy of Violence against Women*. Oxford: Oxford University Press.

Truth and Reconciliation Commission of Canada. 2015. *Honouring the Truth, Reconciling for the Future*. Ottawa.

Tsetta, S., G. Gibson, L. McDevitt, and S. Plotner. 2005. "Telling a Story of Change the Dene Way: Indicators for Monitoring in Diamond Impacted Communities." *Pimatisiwin: A Journal of Aboriginal and Indigenous Community Health* 3(1): 59–69.

Usher, P., G. Duhaime, and E. Searles. 2003. "The Household as an Economic Unit in Arctic Aboriginal Communities, and Its Measurement by Means of a Comprehensive Survey." *Social Indicators Research* 61(2): 175–202. https://doi.org/10.1023/A:1021344707027.

Veracini, L. 2010. *Settler Colonialism*. Basingstoke: Palgrave Macmillan.

Vosko, L.F. 2000. *Temporary Work: The Gendered Rise of a Precarious Employment Relationship*. Toronto: University of Toronto Press.

– 2002. "The Pasts (and Futures) of Feminist Political Economy in Canada: Reviving the Debate." *Studies in Political Economy* 68: 55–83. https://doi.org/10.1080/19187033.2002.11675191.

Vosko, L.F., V. Preston, and R. Latham. 2014. *Liberating Temporariness?: Migration, Work, and Citizenship in an Age of Insecurity*. Montreal and Kingston: McGill-Queen's University Press.

Walsh, D. 2012. "Using Mobility to Gain Stability: Rural Household Strategies and Outcomes in Long-Distance Labour Mobility." *Journal of Rural and Community Development* 7(3): 123–43.

Watkins, M. 1963. "A Staple Theory of Economic Development." *Canadian Journal of Economics and Political Science* 29(2): 49–73. https://doi.org /10.2307/139461.

Watkins, M., ed. 1977. *Dene Nation: The Colony Within.* Toronto: University of Toronto Press.

Weaver, J., N. Todd, and L. Coates. 2007. *Honouring Resistance: How Women Resist Abuse in Intimate Relationships.* Calgary: Calgary Women's Emergency Shelter.

White, G. 2002. "Treaty Federalism in Northern Canada: Aboriginal– Government Land Claims Boards." *Publius: The Journal of Federalism* 32: 89–114. https://doi.org/10.1093/oxfordjournals.pubjof.a004961.

– 2009. "'Not the Almighty': Evaluating Aboriginal Influence in Northern Claims Boards." *Arctic* 61(5): 71–85. https://doi.org/10.14430/arctic103.

Wilson, J. 2009. *Poverty Reduction Policies and Programs, Northwest Territories.* Social Development Report Series. Commissioned by the Canadian Council on Social Development.

Wolf, P.R., and J.A. Rickard. 2003. "Talking Circles: A Native American Approach to Experiential Learning." *Journal of Multicultural Counselling and Development* 31(1): 39. https://doi.org/10.1002/j.2161-1912.2003.tb00529.x.

Wolfe, P. 2006. "Settler Colonialism and the Elimination of the Native." *Journal of Genocide Research* 8(4): 387–409. https://doi.org/10.1080 /14623520601056240.

Wotherspoon, T., and V. Satzewich. 2000. *First Nations: Race, Class, and Gender Relations.* Regina: Canadian Plains Research Center.

Wright, M. 1999. "The Dialectics of Still Life: Murder, Women, and Maquiladoras." *Public Culture* 11(3): 453–74. https://doi.org/10.1215 /08992363-11-3-453.

YK Dene First Nation Advisory Council. 1997. *Wıìlıìdeh Yellowknives Dene: A History*, Dettah: Yellowknives Dene First Nations Council.

Young, I. 1980. "Socialist Feminism and the Limits of Dual Systems Theory." *Socialist Review* 50(5): 169–88.

Zoe, J.B. 2005. "Tłı̨chǫ History." *The Tłı̨chǫ History Project.* Website hosted by the Tłı̨chǫ Nation, Behchokǫ̀.

– 2014. "Tłı̨chǫ History: Stories and Legends." *The Tłı̨chǫ History Project.* Website hosted by the Tłı̨chǫ Nation, Behchokǫ̀.

Index

Abele, Frances, 32, 35, 39, 105, 193–4, 234n9
Akaitcho Hall, 68, 70
Akaitcho Nation, 109, 232n7
Alaska, 55
Alberta oil industry, 83
Alfred, Taiaiake, 42
Amnesty International, 92, 178, 180
Anderson, Kim, 178, 195, 224, 241n15
Angola, 84, 87–8, 90, 95, 217
annual Strawberry Ceremonies, 224
anti-racist feminists, 6, 27
anti-racist (theories), 28, 32, 195, 226
anti-violence education initiatives, 227
Arctic, 34, 79, 93, 212, 215
arsenic, 60–1, 100, 213, 237n18
Asch, Michael, 31, 34–5, 51, 235–6n9
Aurora College, 135, 209, 242n9
Australia, 101, 125, 127, 132, 244n7

Banaji, Jairus, 33, 37
Bannerji, Himani, 33, 41
Baskin, Cyndy, 50
Bay Street, 90, 94, 241n2
Beaufort Delta, 188–90, 208
Behchokǫ̀, 7–8, 17, 60, 65–6, 108, 137, 158, 160–1, 163–5, 173, 186, 188, 194, 208, 232n7
Belgium, 88, 91

Bennholdt-Thomsen, Veronika, 31, 183, 198–9, 223, 245n7
Berger Inquiry, 107
Blondin, George, 35, 50, 198, 235n9
blood diamonds. *See* conflict diamonds
Boots on the Ground, 212
boom-and-bust, 97–104; cycle, 97, 100; Dene concept of rupture and, 98; diamond industry and, 96; dynamics, 98; economies, 97; resource extraction and, 31
Bourgeault, Ron, 13, 52, 235–6n9
British imperialism, 86
Broken Hill Proprietary (BHP), 101

Caine, Ken, 115, 213, 241n14
Calder et al. vs. Attorney General of British Columbia, 107
Canadian Broadcasting Corporation (CBC), 18, 63, 237n18, 239n8
Canadian diamond industry, 83; businesses auxiliary to, 119, 124, 239n8; exceptionalism of, 79, 90, 93; gender divisions and, 136; management of, 17, 19, 229; as responsible, 22, 89–90, 93, 95, 97, 218; training for, 134; wage labour and, 124, 145

Canadian diamond-mining regime, 5; approaches to Indigenous identity, 111–12, 120, 220; approaches to Indigenous relations, 22–3, 45, 96, 104–5, 118, 120, 185, 219; emergence of, 111, 120, 151; feminization of social reproduction and, 124, 175; gold mining (*see* Canadian gold-mining regime); Indigenous households and, 159, 176, 204; northern Indigenous organizing and, 97, 100; masculinity and, 180; temporariness of, 46, 104, 201, 219, 227; wage labour and, 113, 123, 136, 159, 163. *See also* FIFO diamond-mining regime; Indigenous women; male-breadwinner/female caregiver model; social relations; social reproduction; subsistence; violence
Canadian gold-mining regime, 22, 57, 156; diamond-mining regime and, 11, 22, 120, 219; end of, 57, 97, 239n1; Indigenous engagement and, 60, 111; temporary permanence of, 219. *See also* Indigenous women
Canol Pipeline, 55
caribou, 9, 35, 64, 235n3; concerns for, 211–13; decline of, 44, 60, 203, 212, 227; diamond mines and, 211–12, 227; displacement of, 201; hunting of, 36, 52, 61, 223; migration disruption, 55, 185, 211. *See also* Indigenous women; subsistence
CBC Radio North, 62
CEDAW (Committee on the Elimination of Discrimination Against Women), 178
Centre for Northern Families, 146
chartered flights, 216
Cherwaty, Mary Lou, 131–2, 242n6

Chief Monfwi, 108
Child and Family Services, 73
children. *See* Indigenous children
Christensen, Julia, 70–1, 73, 138–9, 162
coal-mining town, 135, 175–6
commodity fetishism, 82
Communities and Diamonds Report, 113–14, 185, 187, 189, 190, 245n4
community and kin networks. *See* kin and community networks
Con Mine, 54–9, 98, 156, 236n11, 236–7n15
conflict diamonds, 79, 80, 83–4, 86–90, 95, 98
corporate social responsibility, 94, 141
Coulthard, Glen Sean, 4, 6, 10, 32–3, 41–2, 85, 109, 182, 190, 194, 210, 223, 231n5, 234n2, 235n7
country food, 179, 186, 188, 195, 198, 225, 234n5, 244n2

Davis, Angela, 165
De Beers, 22, 80–3, 86–9, 95, 187, 217
De Beers Group, 82, 101
debt: as dispossession, 159; household, 207; mortgage and, 154, 159, 161; wage labour and, 160–1
Dedats'eetaa, 212
Dene of the Dehcho Region, 51
Deninu Kų́ę́ of Fort Resolution, 109
Department of Indian Affairs and Northern Development, 58
Department of National Defence (DND), 99
Dettah, 7–8, 17, 65–6, 68, 137, 160–1, 186, 194, 232n7
Devolution Agreement, 103
diamond industry (international), 80–1, 83, 86, 89, 90, 93, 95, 217; violence of, 84, 86–8
diamond mines. *See* Canadian diamond industry

diamonds: romantic love and, 22, 81, 83, 217. *See also* Canadian diamond industry; conflict diamonds

Diavik Diamond Mine, 101, 216

dispossession: diamond-mining regime and, 45; as extractive, 22; history of, 8; of Indigenous lands, 54; of Indigenous peoples, 109; settler capitalism and, 7, 235n6; settler colonialism as, 28, 45, 97, 139. *See also* debt

domestic violence. *See* violence

Edmonton, 137, 197, 236n11

education, 125, 134, 136, 141, 153, 197–8, 203, 208–9, 232n10, 238n29; community, 208; intergenerational, 196, 203, 209–10; place-based, 184. *See also* anti-violence education initiatives; subsistence

educational services, 130

Ekati, 92, 101, 104, 114, 116–17, 156, 212, 215, 228, 240n8, 242n5

embodied violence, 20, 213, 243nn16, 1; colonialism and, 44; of diamond mines, 139, 152, 165–6, 184, 224, 226–7; land and, 24, 61, 200, 225; wage labour and, 158, 163. *See also* Indigenous women; violence

Employment and Social Development Canada, 134

entrepreneur, 84, 231n2; Indigenous, 116, 124

environmental assessments, 113, 212

environmental conservation, 64

environmental destruction: cancer and, 213, 244n9 (*see also under* health); flora and, 55, 66; gold mines and, 61, 109, 213; resource extraction and, 155, 179, 200, 210–13, 215; settler land-use and, 60; social impact of, 213

environmental responsibility: diamonds and, 79, 92, 94; Indigenous management and, 50, 108, 221

Eureka diamond, 22, 81, 86

exploitation: capitalist drive for, 98, 100, 118, 199; of labour, 12, 29–30, 34, 75, 98, 126, 166–7, 199, 233, 245n8; of land, 24; of non-capitalist labour, 199–200

extractive labour, 56, 58–9, 85, 124, 137

extractive site, 101, 179, 201

Fanon, Frantz, 42, 44, 226

Federici, Silvia, 33, 167, 176, 243n5

femininity, as untrainable subject, 144

feminist historical materialism, 27

feminized labour, 40, 52, 54, 221

FIFO diamond-mining regime: community organizations and, 136; feminization of social reproduction and, 13, 140; gender relations and, 23, 128, 222; Indigenous participation in, 13, 117; masculinization of production and, 23–4, 123; racialization and, 23; restructuring labour through, 12, 23; temporariness of, 104, 127, 149, 216, 219, 223. *See also* Canadian diamond-mining regime; FIFO model; Indigenous women; male-breadwinner/female caregiver model; social relations; social reproduction; subsistence

FIFO model, 11–12, 19–20, 96, 98–9, 125, 127, 133, 219; detached workers and, 142, 144, 149; flexible labour and, 12, 23, 124, 127; mining towns and, 125–6, 241n3; psycho-social impacts of, 226; schedules under, 165; seasonal

labour and, 191–2; spatializing quality of, 13, 128, 168, 172, 221, 226. *See also* FIFO diamond-mining regime
financial literacy training, 159
Fipke, Chuck, 100–1
fixed capital, 56, 60, 96, 120, 123, 125–6, 215, 219, 241n1
Foreign Investment Protection Agreements (FIPAs), 94
Fort McPherson, 107
Fort Rae. *See* Behchokǫ̀
foster care, 73, 153
Foucault, Michel, 167
fur trade, 13, 34–5, 55, 60, 93, 97, 235–6n9
future generations, 119, 184, 200–1, 215

Galeano, Eduardo, 58, 84, 86
Gatcho Kué, 101, 104, 117, 187, 215
gender relations, patriarchal, 13, 68, 222
Giant Mine, 55–61, 98, 156, 237n18, 245n11
Gibson MacDonald, Ginger, 9, 45, 203
Global Witness, 87
GNWT Diamond Policy, 92
GNWT Legacy and Hope Foundation, 68
gold-mining regime. *See* Canadian gold-mining regime
government housing, 66, 160–1
government of NWT (GNWT), 92, 129, 131, 133, 135–7, 187, 239n3, 245n4; housing and, 138, 161 (*see also* government housing); Indigenous self-determination and, 45; land claims and, 109; mining revenue and, 102; royalties and, 103–4, 112; SEAs and, 22, 113–14, 185–6. *See also* subsistence

government provision, 158
government schools, 64, 67–8, 137
Great Bear Lake, 56
Great Slave Lake, 7, 9, 51, 61, 123
Green, Julie, 135
grounded normativity, 4, 10, 31, 49–50, 85, 109, 179, 182–3, 196
Gwich'in (peoples), 108, 111, 240n6

Hall, Stuart, 42
Harper, Stephen, 92, 94, 231n2, 240n10
health: emotional, 180; mental, 44; perspective, 44; public, 44, 185; spiritual, 212; standards, 127
health care, 37–8, 130; Indigenous delivery of, 108
health concerns, 162, 211, 213, 232n10; cancer and, 44, 180; colonialism and, 44; country food and, 180; debt and, 160; diamond-mining and, 184–5; discrimination and, 143; feminized labour and, 135; extraction and, 179, 244n9; land and, 210, 212, 227; mine workers and, 57; mining sites and, 40, 143, 149; racism and, 149. *See also* well-being
high-wage employment, 57, 98, 124, 131, 138; impact on well-being, 158; Indigenous labour and, 113
homelessness, 70, 138–9
homicide, 178
hooks, bell, 165
household, 5, 35; Canadian state and, 66–7, 73–4, 151; capitalist production and, 195, 199, 202; diamond-mining and, 137, 142–6, 157–60, 171–2, 176, 202, 204–5; expenses of, 152; feminized responsibility for, 13, 56, 128, 136, 140, 142, 150, 163, 169–73, 176, 222;

income of, 137, 158–60, 163, 202;
nuclear concept of, 37, 52, 66, 155,
162, 171, 174; wage labour and,
136, 159, 161, 163–4, 198. *See also*
social reproduction, subsistence

imagined frontier, 56, 179, 231n3
imagined national citizen, 58
immigrant, 204–5; as category, 58;
labour, 58; settlers racialized as, 58
Impact Benefit Agreements (IBAs),
22, 97, 112, 114–17, 132, 203, 220,
240n6, 240–1n13, 241n14
Indigenous children: apprehension
of, 73, 151, 153, 221; Canadian
state and, 64–8, 72, 151; care for,
36, 74, 87, 128, 149, 152–3, 161,
167, 169, 204; community and, 67,
69–70; feminized responsibility
of, 37, 67, 149, 169, 170, 206. *See
also* residential schools; social
reproduction; subsistence
Indigenous organizing, 11, 63, 97,
105–6, 111, 120
Indigenous motherhood, 50, 67
Indigenous/settler relations, 12, 28,
112, 114, 185, 202
Indigenous subjectivities, 28; as
racialized, 42, 112, 220; state
approaches to, 28, 44; and
subsistence, 42
Indigenous women: Canadian
diamond-mining regime and,
123–4, 175–6, 220; Canadian gold-
mining regime and, 139; caribou
and, 52, 211; embodied violence
against, 24, 43, 146, 149, 156, 200,
220–1, 225; FIFO model and, 23,
123–4, 139–49, 169, 174; intergenerational social reproduction,
13, 20; negotiating ruptures, 155,
156, 171, 200; racism and, 43, 143,

197, 200; resistance of, 46, 151–2,
155, 183–4, 200, 202, 206, 210–11,
214, 222–3, 227; restructuring of,
labour, 24, 74; surveillance of, 74,
221, 244n8; violence against, 5, 21,
43, 149, 156, 172, 176, 178, 218 (*see
also* violence); well-being of, 17,
149, 152 (*see also* health)
industrialization, 144, 179
Innis, Harold, 93
intergenerational social
reproduction, 52–3, 57, 120, 137,
206, 211–12, 214, 221–3; capitalism
and, 119; collective wellbeing
and, 11, 28; of community, 24,
38, 49, 71, 123, 128, 151, 163, 206,
213; education and, 69, 196, 203,
207, 209–10, 222; knowledge
transmission and, 151, 183, 196,
209–10; of labour power, 40, 166;
place-based ontologies and, 104,
155, 227; residential schools and,
70; workers and, 56, 169. *See
also* Indigenous women; social
reproduction; subsistence
inter-household relations, 154–5,
171, 184, 195, 203, 205
inter-household networks, 195. *See
also* kin and community networks
International Union of Mine, Mill
and Smelter Workers, 57
Inuit elders, 212
Inuvialuit, 106, 108, 240n6
Inuvialuit Final Agreement, 108

Jacobs, Erasmus, 81
Jean Marie River, 186

Kakfwi, Stephen, 69, 91, 209
Kimberley Process, 86, 88, 90–1, 95
Kimberley, South Africa, 88
Kimberlite deposits, 22

kin and community networks, 13, 37, 162, 164, 195, 204–5, 209, 222
Kino-nda-niimi Collective, 40, 44
Klondike Gold Rush, 97
Kuokkanen, Rauna, 174, 196, 200

land-based activities, 4–5, 11, 12, 28, 38–9, 49, 183, 192, 223
land-based economies, 34, 47, 49–53, 69
land-based production, 7, 31, 54, 178
land-based relations, 24, 49, 66, 111, 120, 128, 179, 182–3, 201; reproduction of, 31
land claims, 103, 107; Canadian state and, 103, 109–10, 240n11; cultural recognition and, 32; Dene Nation, 109; diamond-mining regime and, 111–12; Indigenous organizing and, 111; Indigenous resistance and, 63; Indigenous self-determination and, 108, 110; negotiation and, 32; northern, 108–10; subsurface rights and, 109; Tłı̨chǫ, 7, 108. *See also* Tłı̨chǫ Land Claims and Self-Government Agreement
land tenure, 42, 110, 240n10
Latin America, 84, 180
Lawrence, Bonita, 52, 195, 224
lean operations, 125–6
Little, Lois, 107, 135, 209
Łutselk'e, 109, 240n13
luxury minerals, 86, 89

MacDonald, John A., 67
Mackenzie Valley, 107, 111
Mackenzie Valley Gas Pipeline, 106, 117
Mackenzie Valley Pipeline Inquiry, 63
male-breadwinner/female-caregiver model, 56, 136, 154–5, 174,
214; Canadian state and, 206; diamond-mining regime, 67, 155, 171–2, 175, 178, 202, 221; gold mining and, 57
Marx, Karl, 29–30, 33–4, 36, 41, 82, 85, 126, 166–7, 233nn1, 2, 3, 243n4
master's tools, 106
metabolic rift, 85
metropole, 101
Mies, Maria, 39, 183, 198–200, 223, 243nn5, 7
Mineral Resources Act, 104
mineral royalties, GNWT, 103
mines. *See* Canadian diamond-mining regime; Canadian gold-mining regime; FIFO diamond-mining regime
mining camps. *See* FIFO model: mining towns and
mining industry. *See* Canadian diamond-mining regime; Canadian gold-mining regime; FIFO diamond-mining regime
missionary education, 52
modernization, 93
modes of production, 34, 36; place-based, 111, 156; settler colonialism and, 217; temporal, 156
Mountain Province Diamonds, 101, 104

Nahanni, Phoebe, 51–2, 107, 174, 209
Nanasivik, 127
National Day for Murdered and Missing Indigenous Women, 224
National Inquiry into Missing and Murdered Indigenous Women, 172, 178
Native Women's Association of the NWT, 5, 15, 232n11
Natural Resources Canada, 91
Ndilo, 188–90, 194

needs of capital, 41, 66, 126
neoliberal industrial regimes, 12
neoliberal restructuring, 100
networks. *See* kin and community
   networks
NGOs, 86–90, 95, 156, 218
nomadic lifestyle, 35, 51
nomadic patterns, 56, 64–5, 191
Norman Wells, 55, 242n11
northern autonomy, 45, 103, 234n8
northern environmental programs, 64
northern settler administration, 101
Northern Territories Federation of
   Labour, 131
Northern Women in Mining, Oil and
   Gas Project, 134
nuclear family, 30, 66, 83, 155, 181,
   195, 221
NWT Federation of Labour, 19, 131
NWT Housing Commission, 160
NWT Mine Training Society, 134
NWT and Nunavut Chamber of
   Mines, 19, 101, 103, 112, 131,
   134, 141
NWT Union of Northern Workers, 57

Office of Native Claims, 107
Ontario, 5, 14, 62, 90, 99
ontologies (Indigenous), land-based,
   38–9, 42, 110–11, 189, 224, 227
Ottawa, 5, 218

Partnership Africa Canada (PAC), 87
patriarchy, 41; as ideology, 206, 214;
   patriarchal state, 243n5 (*see also*
   gender relations; nuclear family);
   power relations and, 52–3
Polaris, 127
post-diamond: future, 104; mixed-
   economy, 227–8
poverty, 56, 65, 74, 85, 153, 158; in
   NWT, 137, 177; punishing of, 74;

racialization of, 74, 139; as social
   problem, 114
pre-capitalist magic, 167
prejudice, 43, 109
private housing, 159
public housing, 138, 159–62

racism, 42–3; of capitalism,
   168; colonial, 69–74, 86, 168;
   commodity, 93; healing from,
   197; mining sites and, 143, 173;
   violence and, 200, 225–6. *See also*
   Indigenous women
Razack, Sherene, 178
real abstraction, 167
relocation initiatives, 64
research methods, 14–20
residential schools, 15, 44, 67–74,
   106, 208–9, 221, 225, 238n28
resilience, 61; of community and
   kin relations, 202, 204–6, 222;
   mixed economy and, 74, 225; as
   rhetorical device, 245n9; social
   restructuring and, 205
resistance, 24, 46, 63, 105, 107, 118,
   182, 224, 226; acts of decolonizing,
   20, 27, 210, 222; beading and
   sewing as, 151–2; to colonization,
   4, 27; community and kin
   relations, 118, 202–3, 206, 222–3,
   243n14; to extraction, 46; feminist,
   anti-capitalist, 198; mixed
   economy as evidence of, 10, 31,
   181; reorganizing labour-power
   and, 40. *See also under* Indigenous
   women; social reproduction
responsibility paradox, 94
Revolutionary United Front, 87
Rhodes, Cecil, 81, 86
Rio Tinto, 101, 216
Royal Canadian Mounted Police
   (RCMP), 65, 90, 99, 127, 246n1

Royal Oaks, 98
rupture: diamond-mining regime and, 10, 31, 61, 75, 98, 104, 120, 137, 151, 156, 171, 183–5, 215, 218, 227–8; mixed economy and, 10, 31, 55, 97, 226; settler colonial violence and, 227; state-driven, 165, 221; Tłı̨chǫ concept of, 9, 10, 45, 98, 157, 184, 199. *See also* Indigenous women; social relations; social reproduction

Sahtu, 108, 111, 188–90, 193–4, 199, 245nn5, 6
Scott, Rebecca, 135–6, 175–6, 227
scramble for Africa, 86
Seccombe, Wally, 32–3, 233n4
Second World War, 35, 54–5, 103, 125, 231n2, 235n8
self-harm, 176
sexual harassment, 142–3, 146–9
Sierra Leone, 84, 87–8, 95, 217
silver, 56, 83–4
Simpson, Audra, 43, 67, 200
Simpson, Leanne Betasamoksake, 10, 31, 49, 179, 182, 223
single parenthood, 168
Sisters in Spirit, 178, 224
small local communities (SLC), 133
small social work program, 135
Smillie, Ian, 81, 87–8, 91, 239n6
Snap Lake diamond mine, 101, 187, 227
Speakman, Marie, 43
socio-economic agreements (SEAs), 22, 97, 112–15, 117, 132, 185–6, 220
social relations: Canadian diamond-mining regime and ruptures to, 5, 13, 181, 228; FIFO diamond-mining regime and ruptures to, 125, 128, 185, 191; intergenerational social

reproduction and, 190, 202, 210; ruptures to, 9, 22, 45, 55, 57, 157, 199, 201, 222; violence and restructuring of, 40, 152, 155, 228
social reproduction: Canadian diamond-mining regime and ruptures to, 151–2, 155, 157, 181, 183, 204, 221; FIFO diamond-mining regime and ruptures to, 13, 153, 155–6, 169, 181, 184, 204, 220–2; of households, 13, 23, 30, 142; of Indigenous children, 37, 167; as site of resistance, 14, 36, 155, 181, 206, 214, 218, 223; as site of rupture, 137, 151, 155, 157, 171, 221; violence of separation of production and, 13, 27, 40, 154, 167–8, 176, 200, 221, 236n12
social services, 64, 66, 73, 103, 108, 162–4, 232n10, 244n8
South Africa, 22, 81, 86, 88, 217
southern Canada, 8, 11, 56, 58, 63, 68, 73, 135, 142, 144, 203, 240n12
sphere of labour, 41
Stanton Territorial Hospital, 156
Starblanket, Gina, 52–3
Status of Women Council of the NWT, 141
Subarctic, 34, 90, 128, 212
subsidized housing, 66, 160, 237n25
subsistence: Canadian diamond-mining regime and ruptures to, 11, 105, 124, 156, 183, 185, 191, 200–1, 223; caribou as, 52, 195, 203, 211, 223; education, 197, 207–10, 222–3; FIFO diamond-mining regime and, 24, 140, 183–5, 190–3, 202, 223; government of NWT and, 184–6; households and, 49, 163, 184, 187, 191, 193–5, 199, 202, 223; Indigenous children and, 51, 196, 214; intergenerational social

reproduction and, 14, 156, 183–4, 188, 196–7; well-being and, 40, 195
subsistence perspective, 183, 198, 223
substance use, 44, 164–6, 244n7
subsurface rights, 62, 237n22; federal government and, 102; Indigenous claims to, 110; land claims and, 109
suicide, 177–8, 189, 202
supra-household interaction, 195
surplus-value, 241n1; accumulation of, 12, 29, 31; creation of, 12, 34, 36, 166–7; extraction of, through labour, 4, 11, 105, 179, 182; production of, 183, 233n1
surveillance, state, 67, 73, 221, 245n10; of Indigenous households, 73. *See also* Indigenous motherhood; Indigenous women

temporary foreign worker program, 8, 136, 139, 232n9
Thobani, Sunera, 66–7, 221
Tłı̨chǫ (peoples), 8, 34, 50–1, 54; approaches to relationships, 10, 45; communities, 7; cosmology of, 9, 45; diamond mines and, 116, 118; Elders, 8, 45; fur trade and, 60; gold mines and, 56, 59–60, 100; history of, 9–10, 28, 45; land claim and, 108 (*see also* Tłı̨chǫ Land Claims and Self-Government Agreement); land use and, 51; language and culture of, 8; Logistics, 118; relationships and, 9; resource revenue and, 111; summer camps of, 9
Tłı̨chǫ Land Claims and Self-Government Agreement, 7, 108, 110
tourism, 9, 131
traditional hunting camps, 64–5
transnational capital, 12
transnational labour, 58
trap line, 9, 179, 212–13

Treaty 11, 108
treaty payments, 65–6, 137, 163
Trudeau, Justin, 92–3
Truth and Reconciliation Commission, 69

union organizing, 126, 131–2, 236n14
unions, 57, 98–9, 148, 219
UNITA, 87
uranium, 55–6, 97, 244n9

Veracini, Lorenzo, 220
violence: of diamond-mining regime, 7, 86, 94, 139, 165, 180, 184, 202, 218, 224, 226–7; extraction and, 22, 24, 61, 83–5, 87, 89, 95, 98, 179–80; income and, 158, 163; intimate partner, 5, 178, 225, 246n9; land-based, 24, 179, 210, 225; at mining sites, 124–5, 146, 148–50, 220, 225; reorganizing labour and, 21, 24, 27, 40, 149, 180, 224; of settler-colonialism, 13, 21, 38, 43–5, 71, 178, 200, 202, 218, 225; of sexual harassment, 40, 43, 125, 142–3, 146–9, 153–4, 178, 225; substance use and, 164–6, 244n7. *See also* Indigenous women; social relations; social reproduction

wage labour, 9, 129, 168; gold mining and, 63, 156; household (*see under* household); as masculinized, 136; mining (*see under* Canadian diamond-mining regime); mixed economy and, 9, 63, 136; "necessity" of, 118, 133, 154, 159–61; non-mining related, 129
water, 9, 48, 50–1, 108, 124, 127, 238n4; as commodity, 215; contaminated, 61, 215; de-watering process and, 180; diamond mines

and, 201, 213; as life, 228; pipes and, 162; relationships to, 64

Watkins, Mel, 35, 93, 107

welfare: payments, 64–5, 151; policies, 65–6; state, 65–6

well-being, 39, 103, 110, 118, 158, 203, 207, 210; cultural, 113, 186; collective, 11, 28, 195, 198; diamond mines and, 178, 185, 232–3n11; emotional, 39–40, 158, 178; family, 152–3, 157; land and, 180, 210, 212, 224, 227; material, 110, 178; spiritual, 110. *See also* health concerns; Indigenous women; subsistence

White Paper, 1969, 107, 240n9

Whitehorse, 55

Wıìlıìdeh (peoples): diamond mines and, 54; disease and, 54; gender relations of, 49; gold mines and,

62; land use patterns, 51, 64; mixed economy of, 27, 48–75; summer camp, 9; traditional economy, 34; Yellowknife River and, 50. *See also* Dettah; N'dilo

Wolfe, Patrick, 42, 11

women. *See* Indigenous women

worker: as detachable, 144; as low-skilled, 144–5; untrainable, 124, 144–5, 243n15

workspace, 126, 156

world systems (theory), 98

Wright, Melissa, 124, 144–5, 243n15

Yellowknife River, 9, 50–1, 60

Yellowknife's Old Town, 7, 55

Yellowknives Dene First Nation Elders Advisory Council, 64–5, 236n9

Zoe, John B., 7, 9, 108, 227

www.ingramcontent.com/pod-product-compliance
Lightning Source LLC
Chambersburg PA
CBHW030239030426
42336CB00009B/163